MODERN BRITISH
FARMING SYSTEMS

MODERN BRITISH FARMING SYSTEMS

(An Introduction)

edited by

Frank H. Garner
Former Principal, Royal Agricultural College,
Cirencester

PAUL ELEK
(Scientific Books) Ltd.

Copyright © *Frank H. Garner and others 1972*

Published in Great Britain by
Paul Elek (Scientific Books) Limited
54–58 Caledonian Road, London N1 9RN

Library of Congress Catalog Card Number 79–80930

ISBN 0 236 17730 3

Printed in Great Britain by
The Garden City Press Limited
Letchworth, Hertfordshire SG6 1JS

CONTRIBUTORS

Editor: Frank H. Garner, MA, MSC, FRAGS., Former Principal, Royal Agricultural College, Cirencester, Glos., GL7 6JS, England

D. M. Barling, MSC
H. Catling, NDAGRE, AMIAGRE, AMASAE
G. A. Cragghill, BSC
J. A. R. Lockhart, BAGR
R. G. Norman, BSC, DIPFBA
A. J. L. Wiseman, NDA, CDA
(All the above are members of the staff of the Royal Agricultural College, Cirencester, Glos., GL7 6JS, England)
Revd D. G. Peck, Shellingford Rectory, Faringdon, Berks., England

CONTENTS

	page
Preface	
1. A short introduction to the history of British agriculture	1
2. Introduction to farming systems	17
3. All-arable systems	35
4. All-grass systems	59
5. Mixed arable and grass systems	72
6. Livestock production on hill and upland farms	94
7. Intensive livestock systems	104
8. Fruit, flowers and vegetables	124
9. Breeding of farm livestock	148
10. The role of the plant breeder	157
11. Farm mechanisation	168
12. Economics of British farming systems	200
List of books and references	240
Index	243

PREFACE

After many years' experience of teaching students, I have long felt that an introduction into agriculture should embrace an explanation of why certain types of agricultural enterprise are practised in certain areas, and in some instances only in those areas, whereas other types of enterprise are to be found more or less everywhere. It is essential for students to analyse the reasons for these differences, and it is hoped that this book will stimulate interest and encourage them to do so.

This book has been written for all students of agriculture, whether they are attending Farm Institutes, Colleges of Agriculture (as many of the former are now called), the National Colleges of Agriculture, or the many courses at the Universities where agriculture is included. It should also appeal to students of Geography and Economics when commencing their respective studies. Others who are not taking regular courses, but who wish to study agriculture, should also find this book of interest to them.

For hundreds, even thousands, of years food has been produced in the United Kingdon by systematic farming. Initially the food was for local consumption only, but as the population gradually congregated in towns, and the means of communication improved, food was produced in those areas where it could be produced most easily and cheaply. All will accept this as sound policy; in fact it is the agreed policy of the members of the European Community.

Food may be produced in various forms. In some cases it is directly ready for human consumption: examples are oats, or milk or many fruits and vegetables. Or it may be that the raw product which is grown must be processed before it is consumed by human beings, as are wheat for flour, barley for beer, and sugar beet for sugar—others might be quoted. Some of the foods from the farm are really for livestock consumption before the human food is produced; examples that readily come to mind are the cereals which figure so prominently in the feeding of pigs, poultry, cattle and sheep.

So far no mention has been made of the food most commonly grown for livestock, namely grass. In summer, this is the basic food for most cattle and sheep (and also horses where they are still to be found on farms). In winter, grass is used as hay, dried grass and in a wetter form as silage. At the present time grass has not been so prominent in the feeding of pigs and poultry; but there is just the chance that when the United Kingdom enters the Common Market, cereals may become more expensive; this may focus attention upon the greatest use of grass in its various forms for all farm animals.

Since the various types of farming enterprises are dependent upon local knowledge of the soils and climate, it has been found essential to enlist the assistance of experts with specialist knowledge. This has been achieved by invoking the help of various members of the teaching staff of the Royal Agricultural College, in some cases to write chapters or parts of chapters, and also at times for advice on various technical matters. These contributors are listed elsewhere.

Thanks are therefore due, not only to the members of the Royal Agricultural College staff for their constructive help and criticisms, but also to various members of the staff of the Ministry of Agriculture, Fisheries and Food for their willing assistance; and also for permission to use other data, which have been duly acknowledged. Great help was readily given by the photographic staff of the *Farmer and Stockbreeder* who placed their library at the disposal of the authors. The aerial photographs were made available by the willing help of Aerofilms Limited, whose patience in allowing photographs to be borrowed for long periods made the preparation of this book as easy as possible. Finally thanks are due to the Publishers for their indulgence in meeting the needs of the various authors.

Aylesbury   Frank H. Garner
March 1972

Chapter 1

A SHORT INTRODUCTION TO THE HISTORY OF BRITISH AGRICULTURE

This short sketch of the agrarian history of England must start with two explanations. It begins with medieval England. The Normans conquered a land already old, with the techniques and basic organisation of agriculture that were to prevail for the next centuries already developed. The chronology of agricultural development in Saxon England is, however, obscure and the evidence fragmentary. It seemed best for present purposes to ignore discussions of such origins of medieval farming and, indeed, of the fascinating but even more fragmentary evidence of still earlier farming, and to begin at the point where the survival of evidence makes possible a reasonably continuous history of the rural economy. Secondly, much work has been done in the last two decades or so on agrarian history—witness the appearance of the first volume of a comprehensive *Agrarian History of England and Wales* (Vol. IV, in fact, 1500–1640). Much of this work has been done within the discipline and with the tools of economic history and the emphasis has been on exploring the interplay of agricultural and other social and economic developments. It is as an introduction to such work that this essay has been planned.*

THE MIDDLE AGES

Accounts of medieval agrarian life are usually constructed round the "manor", and this is still a reasonable framework; though few historians would care to resurrect the once familiar "model" of manorial farming as an adequate and universal account of agriculture in the Middle Ages. The

* A selected bibliography on pages 15 and 16 includes works to which the author is indebted.

component parts of these estates were the demesne (the lord's land); the land in the occupation of tenants of varying degrees of servitude, mainly holding by servile tenures "in villeinage"; a dependent peasantry; and rights of lordship exercised by the lord over the inhabitants. There certainly existed manors on which the lord exercised his lordship powerfully over unfree tenants, who were bound to the soil and cultivated a demesne occupying over 30 per cent of the available arable by "week" and "boon" work. Yet the manorial organisation was capable of great variation from place to place and even in the same place from time to time. The weight of lordship varied but a more important clue to variety is in the size and the use made of the demesne. Some manors had little demesne and labour services were correspondingly light. Demesnes were sometimes leased out, not infrequently to the villeins themselves.

We can do little more here than indicate the main course of change in the Middle Ages. To this end, certain misconceptions must be dismissed. The course of change was not simply from less to more economic freedom and independence for the peasantry. The manorial economy was not "natural", self-contained and self-sufficient. The need to find rent and a multitude of dues must have dominated peasant farming, while the lords' needs for cash certainly affected their estate policies. Thirdly, the villeins were not economically a homogeneous class. There were substantial farmers among them and they themselves could be employers of labour. Freedom in status, moreover, was no guarantee of economic independence.

The course of change until about 1300 was dominated by population growth. This led to an extension of settlement and to an extension of arable in old settlements—but a limit to this was reached. Indeed, if the evidence of Domesday as interpreted by Lennard and others is to be believed, there was already in the 11th century an astonishing amount of land under arable cultivation—a situation made necessary by low yields and fallows. This trend led to pressure on grazing land and a more stringent control of common grazings and to a dangerously low level of animal husbandry. Extension failed to meet the demand for land, and holdings began to fragment—to such an extent that many could not have supported a family. One random sample of 13th-century manors showed that 45 per cent of the inhabitants had less than five acres apiece.

Against this background of poverty, land hunger, cheap labour and rising prices, the ebullient demesne farming of the 13th century took shape. The trend then was to direct exploitation, using a good deal of paid labour. Grain and wool sales played an increasing part in estate incomes, though rent continued to provide a sizeable proportion. The

13th century was a "boom" period which those situated advantageously could ride. By 1300, however, the agricultural base of the boom was disappearing. There is evidence of recession, of retreat from marginal lands. Land was "de-colonised" and the movement of expansion went into reverse. There is also some evidence of soil exhaustion, with exhausted land being let off to peasantry still land-hungry. Late 13th-century England has been described as "densely occupied, predominantly corn growing and deficient in pasture".

This situation invited the disasters which befell in the 14th century. Terrible famines marked the second decade, and in the middle of the century came the Black Death which wiped out (with later visitations) between a third and a half of the population. This at least relieved the pressure, and in fact established a quite different economic climate. Where its effects were not very immediate and where the manorial "system" apparently continued unchanged, this was because a considerable reservoir of landless labour remained to be upgraded to villein tenancies. The numbers of small proprietors with less than five acres dropped sharply.

The supports of demesne farming, the buoyant market and the cheap labour, had vanished. Profitability had gone, and the landlords learned to settle for a rentier position as best they could with the peasants often in a strong bargaining position. Landlords alienated demesnes, letting them out on leases, often enough to the villeins or consortiums of the wealthier ones. Labour dues had thus no further meaning, and the villein holdings also went for money rents. They were held by copy of the agreement on the roll of the manor court—copyhold. But not all of them—some villein land had fallen in hand through lack of takers, and this might eventually be subject to some form of leasehold.

There is a very different atmosphere about the 15th century, but it is difficult to assess the trend of the rural economy. Land was plentiful and rents were low. Wages remained high. There is evidence that relief from the compulsion to grow grain had beneficial results in the form of more varied agriculture—more legumes, more pasture, more animal husbandry. More barley seems to have been grown too—the demand for beer being more elastic than that for bread—while constant wheat prices suggest that the demand for bread grains was comfortably met. In the absence of farming by the great estates, however, evidence is tantalisingly scanty, for those who undertook the farming were not the kind to keep records. Some historians have considered the period one of decay and stagnation: others suppose it to have been one of a modestly prosperous stability with the economy able to take care of a population only about half that of 1300; but we shall never be sure whether the 15th century was or was not "the golden age of the English peasantry".

THE SIXTEENTH CENTURY: STIMULUS AND DEVELOPMENT

The agrarian history of the 16th century—and until 1640—used to be dominated by a discussion of "the agrarian problem" of rackrenting, enclosure (supposedly for sheep ranching), and depopulated villages. This emphasis missed the significance of agrarian change in this period. Lying behind the changes in land use and organisation was the continuing, if uneven, price rise which multiplied the price of agricultural products by more than 6. This inflation is no longer regarded as of purely monetary origin and caused by the debasement of the coinage and the influx of silver, but as having some "real" cause in a substantial population rise.

Whatever the cause, landlords whose income derived from fixed rents—from the Crown downwards—were embarrassed by the rise and took steps to counter it. On leasehold properties, rents could sooner or later be raised, and they were; but on old villein land, and occasionally on old demesne, varieties of copyhold prevailed. Tenurial arrangements in the 16th century were complicated, but the picture that emerges from a study of them does not suggest a wholesale, unlawful dispropriation of the peasantry. Certainly some landlords put on the pressure and some no doubt followed Fitzherbert's advice (*c.* 1530) to convert copyhold to leasehold for lives where possible "for a set day cometh at last". But increased rents and entry fines were paid, and it is more interesting to discover how this was done than to note the grousing which accompanied it.

Enclosure has been another bugbear to historians of this period, but as Dr. Joan Thirsk has shown in what is the most complete and up-to-date discussion of the subject (*Agrarian History of England and Wales* Vol. IV), enclosure in the 16th century was a portmanteau word for a variety of changes in land use in which all classes were concerned. There was certainly some large-scale depopulating enclosure for sheep walks, though these are not easily dated with precision and did not occur after the middle of the century when the collapse of the Antwerp market made it difficult for English merchants to find a "vent" for cloth. In fact, there is evidence for the abandonment of these large-scale enterprises after 1550. Everywhere, however, there were enclosures, often by small men, which did not necessarily amount to very much in any one place or at any one time in terms of acreage enclosed or messuages destroyed, but which did, over the century, mark the increasing tempo of a revolution in agricultural techniques that was working its way over Tudor England.

This revolution, stimulated by the growth of markets and prices, enabled the rents to be paid, and financed "the rebuilding of rural England, 1570–1640", described by W. G. Hoskins. Some of the social consequences of these changes did indeed provoke a contemporary literary uproar as well as some ill-directed and ineffective legislation, but we must not be misled by this. The problem of the 16th century seems not to have been tenurial insecurity but, as so often, the viability of those farmers less well-endowed, equipped or situated in the face of economic change.

The significant changes were not from immemorial arable into permanent grass, but from both arable and permanent grass into "up and down", the 16th-century term for convertible husbandry. The holding of land in "closes" which could be switched to "up" or "down" had been recommended by Fitzherbert, and the practice grew. In this early ley farming, the leys were often long, but the tentative move into mixed husbandry yielded results. Corn yields were increasing and more beasts were kept.

How far this movement had spread by 1640 is not clear. Market stimulus did not operate equally everywhere, but where it did, even the open field system showed itself adaptable to new techniques and the production of fodder crops. The stimulus of cost (rent) was also uneven. Barriers to increased production existed, and a good deal of England remained to be affected; but escape from the shackles imposed on medieval husbandry by the paucity of winter keep was now possible—was, in fact, happening.

According to Kerridge, the peak period of conversion of permanent grass to "up and down" was from 1590–1660. This was the hey-day of woading—woad being a crop useful in taking the rankness out of old pasture land. The fodder-providing technique of floating the water meadows can also be dated into the 16th century. Grazing and hay were improved, but the new fodder crops appeared more regularly in the 17th century. Clover, the spread of which is not to be calculated from imports of the seed, was clipping two-thirds off the price of meadow hay by about 1680, and to be "in clover" was proverbial by at least 1710. Sainfoin, coleseed, rape and lucerne were all grown and turnips were an established field crop in Suffolk by the mid-17th century.

THE EIGHTEENTH CENTURY: THE CRITICAL PERIOD OF INNOVATION

The new crops generated a variety of new systems. At this stage—from about 1680–1740—we enter a crucial period. Inflation was succeeded by a long-run downturn in grain prices, and population growth, too, seems to have slackened. These were phenomena common to Europe, but the responses to these changes were, in England, markedly different to those on the continent. There the natural response was a shift into pasture farming. In England the response was more complex. Some areas were indeed grassed down, but grain production nevertheless increased and England moved into a period of grain surpluses. It has been argued convincingly that this increase in grain production on a falling level of prices was coming from the free-draining light lands where new techniques allowed continuous production with mixed husbandry. Such systems inevitably produced grain, and produced it at a lower cost than could be matched by the traditional grain-growing heavy lands which had a limited acceptance of these new crops and techniques. Moreover, these farms on the light lands, often the "waste" of previous centuries, were the result of enclosure and were privately occupied and farmed by substantial tenants; in contrast to the heavy Midland soils, farmed by small tenants and proprietors, to a large extent still in the open field system.

A number of local studies add point to this argument. Thus in his *History of Broadwell, Oxfordshire*, the Revd. A. S. T. Fisher unearthed in the manor of Holwell a dispute between two owners who had already virtually divided the parish between them. One party was engaged in ploughing up 400 acres of permanent pasture and sowing it to corn, discovering that he had, as a result, to pay increased tithes to the other. He gave notice that he intended eventually to sow a great part of this ploughland with "St. Foyne, Gt. Glover, Hop Clover, Rye Grass, Lucerne and other seeds of Forreine Grasse". The matter was eventually settled and the enclosure of the village agreed between them in the year 1693.

We now realise that the "agricultural revolution" was not confined to a burst of innovations in the later 18th century to coincide with the industrial revolution. It was a long process, its beginnings far back; but if the process was one of evolution rather than revolution, there could still be a critical decade or so when agricultural change was highly significant for economic development. There has been a growing disposition among agrarian historians to find such a significant moment in evolution in the decades round about 1700 rather than in the late 18th century.

The once-common account of the "agricultural revolution" was constructed round a number of factors. These were: the Norfolk "four course"; Parliamentary enclosures (making the adoption of the four course possible); revolutionary methods in stock breeding (pioneered by Bakewell after 1760); and the injection of capital by improving landlords following the example of Coke of Holkham. More recent writers have however disputed this account. The Norfolk four course seems to have been a convenient shorthand symbol for a host of rotations producing fodder crops and largely dispensing with fallows—and some of these were much earlier than the four course. Nor did radical change in cropping systems await Parliamentary enclosures. Even the open fields could be adapted to new systems, and there had been a great deal of piecemeal enclosure before the middle of the 18th century. Indeed, Parliamentary enclosures seem to have been in some places no more than tidying-up operations. The numbers of small landowners—and the area controlled by them—markedly declined in the early 18th century, but not, incidentally, in the era of Parliamentary enclosure. Capital improvements did not await the arrival of Coke, nor did tenant improvements depend on long leases, of which, anyway, there are early 18th-century examples. Even knowledgeable stock breeding which certainly made advances in the later part of the century did not suddenly appear upon the scene.

Thus, to an important degree, the techniques and the organisation, so often associated with the later part of the century, were present in the early years. The importance of the earlier date for a significant advance in agriculture, which is held to have improved productivity by 25 per cent in the first half of the century, is that it may provide a clue to the role of agriculture in historic economic growth. It would appear that in this period agricultural production went ahead of population growth, creating a reservoir of potential food production which absorbed the demands of a later upswing of the population and its urbanisation. The continued lowness of agricultural prices may have helped, also, to create a domestic market for industrial goods.

All this is not to deny that the traditional period of the agricultural revolution, one of rising prices, increasing population and twenty years of war, saw a very great extension of new methods. Contemporaries might be justified in supposing that this constituted the revolution. *Production*, it has been reckoned, rose 40 per cent, but the rises in *productivity* were confined to the earlier period and it is this that lends it significance.

THE NINETEENTH CENTURY

This rapid extension of innovation provides the clue to understanding the nature of the depression which followed the Napoleonic wars. Agriculture shared in a general recession, and there were problems peculiar to agriculture in such an economic climate. To dismiss the years 1820–1840, however, as a period of unrelieved depression is to leave unanswered a conundrum; for in places successful responses to low prices were evidently made. Production increased, and despite retreat from some hungry marginal land, arable acreage also increased, as did yields. The result was that in 1840, despite a rise of more than 30 per cent, Britain was feeding more of its population than it was in 1811.

The depression was selective rather than universal. Increased production at lower unit cost became a necessary response, and the difficulties of high cost arable farming on the heavy claylands, masked for more than a generation, were critically exposed. The heavy, undrained clays, the traditional "wheat and bean" lands where the three field rotation with a fallow was not uncommon, could not match the mixed husbandry and the variety of rotations on the free-draining light lands, where continuous production was possible. Farming was no longer so limited by the inherent fertility of soils, and many of these light lands could now be managed to produce high yields at relatively low working costs.

"High Farming", in fact, was making considerable strides long before its emergence into full daylight in the third quarter of the century. It was sufficiently advanced for its practitioners to form the English Agricultural Society in 1838, with the motto "Practice with Science". In 1846 the Corn Law of 1815 (modified in 1828) was repealed. Whether it had in fact done much to shelter British agriculture is a debatable point—probably not, at least since the early 1820s, though it may have encouraged some use of marginal land. The withdrawal of protection, however, served to stimulate more agriculturists into following the advice of Caird in his pamphlet, *High Farming the Best Substitute for Protection*. The first generation of Free Trade brought few problems and agriculture enjoyed the prosperity of the "Golden Age".

"High Farming" was a system of "high feeding" (of both crops and stock), and of high inputs to achieve high output. The integration of grain and stock farming in various systems produced higher yields of grain and fodder crops and heavier yields of both meat and milk. The mid-Victorian High Farming was not simply the perfect extractive industry, continually restoring what was extracted; it had begun to take on some characteristics of the "factory" in the sense that it processed inputs from beyond its borders. It depended on imported feeds and increasing use of fertilisers as well as on improved and more varied implements and machinery.

Power did not generally find its way to any very significant extent into cultivation—steam ploughing needed fields of more accommodating size and shape though in some parts of Great Britain, notably in East Anglia and parts of Scotland, it was adopted more widely than elsewhere. Steam power did find its way into barns. Caird noted a Bedfordshire farm where a fixed steam engine saved £200 p.a. in the wages of eight men, cutting the cost of threshing a quarter of wheat from 3s to 8d and being employed for grinders, chaff cutters and root slicers. Farming benefited, however, from steam power in the economies and access to markets provided by the railways and from the addition of another nine millions to a population whose standards of consumption were improving.

As far as developments within agriculture itself were concerned, it was probably drainage, that "master engine of improvement", that mattered most. Thomas Scragg's device for mass-producing Reade's cylindrical clay pipes, together with Fowler's draining plough for laying them, made drainage an economical proposition; there was large-scale investment in this improvement in an attempt to turn the claylands into "turnip farms". How far it succeeded is a matter of debate. It has been argued that "long stranglehold of naked fallows on the claylands" was broken at last, and that mangolds and green crops replaced them. Others have disputed the effectiveness of clayland drainage and consider that the increase in agricultural production continued to come from the ever-improving light lands.

But no matter how technically efficient high farming remained, it ceased to be economically efficient, and the difficulties of sorting out the interlocking enterprises in any accounting system increased the problems of adjustment to change. The facts of this change are well enough known—the arrival of transatlantic grain at prices which British farmers could by no means match, followed by refrigerated meat and dairy produce from New Zealand and Australia exposed

British agriculture to the full meaning of free trade. The "protection of distance" had disappeared and agriculture had to adapt to living in an industrial and commercial state. The impact of these changes was, however, rather more subtle than has sometimes been suggested. T. W. Fletcher has shown that the collapse of grain prices was not wholly disastrous for those whose business was animal products. For one thing, cheap grain was cheap input; for another, it freed purchasing power for their own products. "The kind of man," observed a contemporary, "who had bread and cheese for his dinner now demands a chop." The price fall for animal products (except wool) was modest—and for milk non-existent.

Yet agriculture as a whole suffered a great decline both as an employer and in relative importance as an industry. It made the necessary adjustments. The cereal acreage went down—as did the root acreage—while there was a rise in permanent grass of over a million acres in the grazing districts of England and a million and a half in the arable districts. Some sectors suffered little hardship and not much change; others suffered both, and in terms of human disaster the depression took its toll. But on the eve of the First World War agriculture had settled for a more modest but not altogether uncomfortable existence. Sir Daniel Hall could note that it was "sound and prosperous".

INNOVATIONS IN SCOTLAND AND WALES

Innovation penetrated Scotland slowly and methods of cultivation remained "barbarously simple" in many parts until the 18th century. The inability to produce more winter keep inhibited the development of the pastoral economy, though the cattle trade with England was stimulated by the Union—Defoe remarked on the numbers coming into England and the popularity of the meat of these Scottish "runts". This particular trade was further increased after the "pacification" following the rebellion of 1745. Scots cattle were fattened in the northern counties, in the grazing lands of Leicestershire and Lincolnshire and in East Anglia. The ultimate destination of most was the London market.

Yet, despite the fact that enclosure, in the general absence of common rights and of English forms of landholding, was no problem, and despite the enthusiasm of those who founded the Society of Improvers in 1723, there was little general spread of development from the centres of enlightenment such as the Lothians until 50 years later.

Over much of Scotland successful commercial farming had of necessity to be pastoral. Sheep farming began to rival cattle production in the Highlands at the end of the 18th century as landowners found the improved Cheviots and Blackfaces suitable. These enormous flocks had to winter in the glens which provided the only terrain for subsistence arable farming, and this led to a reorganisation of agriculture in parts of Scotland which had disastrous social results, emptying the Highlands by the "clearances" and emigration and increasing the intractable crofter problem.

Scotland, however, was not given over entirely to pasture farming. A fine mixed farming developed in counties like Aberdeenshire, Forfar and Banffshire, while the large-scale farming of the Lothians became a model of High Farming as soon as the Scots were able to take advantage of new developments. Apart from cattle and sheep, they exported corn and, significantly, estate agents who had a considerable influence on developments south of the Border.

Scotland's influence also spread to many countries abroad. Many of her young people chose to emigrate in the latter part of the 19th century. Their pioneering efforts had a great influence in the shaping of the New World.

Welsh farming has always been confined by the "limiting bleakness" of a country dominated by its great upland core. Only in the more genial and comparatively broad valley bottoms of the Eastern Marches and along the south coast did arable farming make much headway. In these areas some early progress was reported: there was "much clover grass and seed" in Gower as early as 1697. Even on the high uplands a form of burnbaking was practised, and some cereal grown—though this cannot have been very rewarding. Generally, Welsh farming has been, and still is, characterised by small holdings, by the presence of more farmers than labourers in agriculture, and by the predominance of grass over corn.

The notion however that the Welsh uplands, until fairly recent times, produced largely goats is nonsense. The trade in Welsh black cattle, driven on the hoof into England, was always considerable and much earlier than the Scottish cattle trade. The Welsh cattle drover was known in England in the 13th century and was a familiar figure in the 16th. The uplands were also the home of considerable flocks supplying the raw material for the Welsh "cottons"—a coarse woollen cloth—widely manufactured in the northern uplands in the 16th century and onwards.

Over much of Wales, except for the southern coastal fringe and the border lowlands, innovation made but very slow headway. Both opportunity and capital for mixed husbandry were limited, and the meagre use of fodder crops inhibited much expansion of pastoral farming until railways brought in feed to supplement home produced stuffs and lime to sweeten the very generally acid soils. Arable farming made a little growth in the mid-19th century, but entered a downward trend in 1870 while livestock increased rapidly.

Much of Wales is still a country of small hill farms engaged in a pastoral economy, constituting a continuing problem for agricultural policy. The erosion of this sort of farming has been accelerating since the last war, and the uplands of Wales is an area for which an integrated policy of conservation and development will have to be devised if even the present reduced population is to be retained.

THE TWENTIETH CENTURY

By 1914 Britain was a self-supplier of food for weekends only, meeting not a third of the nation's requirements. Moreover, this was done by the most space-consuming methods, so that war brought a need to switch into the most economically produced energy foods which would feed the most people per acre. Delay and dithering in putting these principles into action ended in 1916 when, with the necessary organisation (of which the county executives were the lynch pin) set up, and with the necessary powers of compulsion taken, the Food Production Department launched its plough-up policy with the slogan "Back to the 'seventies". Some preferred other names for the policy, describing it as "food destruction" and its sponsors as "plough maniacs", but it went forward until in 1918 there were under cultivation nearly three million acres more than there had been in 1914. Another million acres of ploughland was scheduled for 1919, but the German spring offensive in 1918 undermined that, and by the late autumn the war was over and the F.D.P. wound up. Agriculture's contribution to victory had been enormous but the costs of all this effort were, as the official historian admitted, "very high".

A guarantee of wheat prices in return for the acceptance of state control had been given in the Corn Production Act of 1917. In 1920 this was replaced by the Agricultural Act purporting to embody the government's permanent peacetime policy. Almost at once, however, the bottom fell out of the grain market because of enormous world surpluses, and the discrepancy between market and guaranteed prices soared to astronomical heights, involving a loss of 18s per quarter on wheat and 19s on oats. With three and a half million acres under these cereals, the government simply abandoned its policy.

Disastrous depression meant a pell-mell retreat into grass and reduced standards in arable farming. The acreage of crops other than grass declined by four million acres, and by 1939 there was a good deal less arable under cultivation than there had been in 1914. The place of agriculture in the national economy further declined. Paradoxically this made it possible to consider assistance. Modest beginnings were made with the British Sugar (Subsidy) Act in 1925, but in the 1930s Britain moved, via the Agricultural Marketing Acts, away from the Free Trade of 80 years. The support policies of the 1930s did not substantially reduce aggregate imports though they gave to them a Commonwealth bias. They did, however, arrest decline and they stabilised prices.

The response of the agricultural industry was variable. Although there was no sudden change from the practices of a sluggish industry engaged in low-cost production, some improvement began to show during the five years or so preceding the Second World War. The Wheat Act, the introduction of the Milk Marketing Boards, the Potato Marketing Board and the Pig Board, and the development of the sugar-beet industry, did much to stabilise the situation. In at least some parts of the country, such as Scotland, agriculture was in many respects well geared to step up production when the need came in 1939.

The story of food production in the Second World War is to some extent a repetition of its story in the first, but it was less of a cliff-hanger. Considering that by 1939 there was much less arable under cultivation than there had been in 1914, the achievements of agriculture in the Second World War were even more remarkable. Arable acreage rose from 12·0 million acres to 19·3 million acres, while permanent grass decreased from 18·8 to 11·7 million acres. Yet though permanent and temporary grass had been reduced by 30 per cent, grazing stock came down by only 1 per cent and the area of grass required to feed a "unit of stock" fell by 30 per cent.

Increase of farm incomes was greatest on the arable farms, both heavy and light, with mixed farms a good second and grassland farms showing the lowest rises. The profitability of arable and mixed farming encouraged some farmers to entertain ideas about the nature of post-war farming which economists found alarming. In fact, however, post-war and world-wide shortages—even of grains—ensured that there would be no repetition of 1920 and even that war-time systems would continue on into the peace. British agriculture was assured of some sort of future and the Agricultural Act of 1947 proclaimed the government's objective of "promoting a stable and efficient industry capable of producing such part of the nation's food as might be desirable in the national interest at minimum prices consistent with proper remuneration" of all concerned. This sounded well—especially as it went with a cautious expansion of agriculture—but it left much undefined, not least the "national interest".

Since the war agriculture has been supported in a number of ways. Though this policy has brought its own problems and the need for continuous adjustment (so that some have wondered whether there has,

Table 1.1
Percentage of selected commodities* produced at home

Commodity	Percentage home produced				
	Pre-war	1946/7	1966/7	1968/9	1971/2
Wheat and flour (wheat equivalent)	23	30	45	43	54
Oats	94	95	98	98	100
Barley	46	96	98	96	90
Potatoes: earlies	n/a	98	64	66	66†
main crop	n/a	99	99	100	98
Butter	9	8	7	11	16
Cheese	24	9	43	41	53
Eggs	61	51	96	96	98
Beef and veal	49	58	73	73	82
Mutton and lamb	36	24	44	39	40
Pork	78	34	99	97	98
Bacon and ham	29	36	33	36	43
Poultry meat	80	72	98	99	99

* Supplied by N.F.U. † Estimated

HISTORY OF BRITISH AGRICULTURE 15

in fact, been a coherent policy), there has been notable progress. Table 1.1 lists a number of food products which are important to the British consumer, and shows that the proportions produced at home rather than abroad have generally been at least maintained, and in some cases increased substantially.

At the time of writing, there are doubts about the effects of the Agricultural Act of 1957. At this point, however, the historian's task is ended.

SELECTED BIBLIOGRAPHY OF MODERN WORKS ON AGRICULTURAL HISTORY

This list is arranged according to the chronology of the subject matter, and is designed mainly for the general student. Many of these works are available in cheap editions.

Reginald Lennard: *Rural England, 1086–1135*. Oxford University Press (1959)

M. M. Postan in *The Cambridge Economic History of Europe*, second edition, Volume I. Chapter vii, Medieval Agrarian Society in its Prime, section 7: England (1966)

J. Z. Titow: *English Rural Society, 1200–1350*. George Allen & Unwin (1969) (see note 1)

R. H. Hilton: *A Medieval Society*. Weidenfeld & Nicholson (1967)

R. H. Hilton: *The Decline of Serfdom in Medieval England*. Macmillan (1969) (see note 2)

Georges Duby: *Medieval Agriculture, 900–1500* in the Fontana Economic History of Europe. Volume 1, section 5 (see note 3)

H. S. Bennett: *Life on the English Manor*. Cambridge University Press (an attractive account which needs supplementing by more recent studies)

Joan Thirsk (ed.): *The Agrarian History of England and Wales, Vol. IV*. Cambridge University Press (1967)

Eric Kerridge: *The Agricultural Revolution*. George Allen & Unwin (1967)

Eric Kerridge: *Agrarian Problems in the 16th Century and After*. George Allen & Unwin (1969) (see note 1)

E. L. Jones (Ed.): *Agriculture and Economic Growth in England, 1650–1815* Methuen (1967)

Paul Bairoch: *Agriculture and the Industrial Revolution, 1700–1914* in the Fontana Economic History of Europe. Volume 3, section 8 (see note 3)

E. L. Jones and S. J. Woolf (Eds.): *Agrarian Change and Economic Development*. Methuen (1969)

E. L. Jones: English and European Agricultural Development, in *The Industrial Revolution*, edited by R. M. Hartwell. Blackwell (1970)

G. E. Mingay: *Enclosure and the Small Farmer in the Age of the Industrial Revolution*. Macmillan (1968) (see note 2)

J. D. Chambers and G. E. Mingay: *The Agricultural Revolution, 1750–1880*. Batsford (1966) (a standard work of interpretation)

E. L. Jones: *The Development of English Agriculture, 1815–1873*. Macmillan (see note 2)

C. S. Orwin and E. H. Whetham: *History of British Agriculture, 1846–1914*. 2nd ed. David and Charles (1972) (a useful text book)

W. E. Minchinton: *Essays in Agrarian History*, Vols. I and II. David and Charles (1968) (a valuable collection of reprinted papers from various sources)

W. G. Hoskins: *The Midland Peasant*. Macmillan (1965) (a most attractive local study of great interest)

W. G. Hoskins: *Provincial England*. Macmillan (1965) (contains essays of interest, including "The Rebuilding of Rural England, 1570–1640" referred to in the text)

Two official histories cover war-time farming in the 20th century.

T. H. Middleton: *Food Productions in War*. In *The Economic and Social History of the World War*. Oxford University Press

K. A. H. Murray: *Agriculture*. In the official history of the Second World War. H.M.S.O. and Longmans

Ernle: *English Farming Past and Present*. Sixth edition, Frank Cass (though now out of date and misleading in some particulars, this is still the only one-volume history covering the whole period. It may still be read with some profit, though the new edition should be used with introductions by G. E. Fussell and O. R. McGregor.)

Note (1) These works are in the series "Historical Problems: Studies and Documents", available as paperbacks.

Note (2) These works are booklets edited for the Economic History Society by M. W. Flinn as "Studies in Economic History". They have useful bibliographies. They are paperbacks.

Note (3) These sections of the Fontana Economic History of Europe appear as separate booklets in paperback form. They carry bibliographies.

Chapter 2

INTRODUCTION TO FARMING SYSTEMS

The production of crops, grass, stock, fruit, flowers and vegetables in any area of the United Kingdom is very much affected by the soils, the climate, the labour available and the returns that can be obtained. The chief exceptions to this general statement are with such stock as pigs and poultry. They are often housed in intensive units (frequently called "factories") where temperature and food supplies are under complete control. It must also be accepted that the modern production of flowers and some vegetables under glass involves growing under strictly controlled conditions.

THE CHOICE OF SYSTEM

The economic returns obtained from various farming enterprises are subject to considerable variations. Some of the farming systems were evolved many years ago when labour was cheap and abundant, using substantial buildings that are difficult to modify to meet the present-day conditions with labour in short supply and expensive. Thus some farmers find themselves wedded to outmoded systems they would change if they had free choice. This is very evident with present economic pressures on sizes of farms. Whereas formerly a farmer could obtain a living from say a 50-acre holding by growing normal crops and by keeping ordinary farm livestock, now it is accepted that at least double this area is essential to make a viable unit. It is now national policy to encourage the enlargement of smallholdings by amalgamations. It must be appreciated that such amalgamations are expensive, since sometimes new buildings may be essential to make viable units for such enterprises as dairying, pig-keeping and grain drying and storing. It may also be essential to amalgamate fields; this may involve modifications of fencing and drainage.

Where the necessary capital is not available to modernise, farmers are forced to continue (with higher labour costs and consequent lower returns) in conditions that are not ideal. They may even be forced to continue with enterprises that are not ideal for the soil and climate because it is too expensive in capital to make a change.

In times of depression when land was cheap it was bought and developed by farmers for systems that were remunerative at that time and these systems may still persist. A very good example to quote is found on the Downs: up to the early 1920s they were used as sheep walks, later they were changed with ley farming into dairy farming with portable milking bails. More recently, after the land had been improved, some fields have been used for almost continuous barley growing while other fields have become permanent grass for dairying with fixed herring-bone parlours.

Finally the actual farming system, though to a large extent influenced by the soil, is developed according to the choice of the farmer to carry out those enterprises in which he is most interested. With the use of modern machinery, manures and sprays it will be seen in subsequent chapters that some farmers have been able to overcome, to some extent, the restrictions that, because of soil, were once imposed upon various aspects of farming. One example is that it is not usual to grow potatoes on heavy clay soils, but with modern equipment and a limited acreage, the tops may be killed by chemical sprays so that the crop may be harvested before the autumn rains make the soil too wet to lift the crop. Many other examples might be given.

With these limitations in mind, the four main factors (namely soils, climate, labour and economics) that influence farming will now be considered in some detail.

SOILS

Examples of most soil types that occur in the world may be found in the United Kingdom; they exist in relatively small pockets and frequently in many different areas. Thus it is quite common to find a large range of different soil types in a small radius, e.g. chalk, gravel, clay and fen and all within 5 miles of Cambridge.

The main soil types may be classified as Clays, Sands, Gravels, Loam, Chalk, Peats, Fens and Mountain Rocks; naturally there are many mixtures of these various basic types. Frequently soil changes follow field boundaries, and this is why boundaries have been so planned. Often it is for this reason that one has very irregular-shaped fields, for it is usually advantageous to have only one soil type in a field. In some cases changes are so sudden that one may have two or more soil types in one field and, arising from this, there may be more than one system of cropping on that field.

Clays

In broad general principles the heavy clay soils are used for production of grass or, if they are under the plough, in the main to grow wheat and beans. Owing to the small sizes of the particles that constitute the soil, drainage can be a major problem, and large sums of money must be spent from time to time to maintain both the drains in the fields and the ditches that run between them. This retention of water is an advantage, for the growing crops can then withstand drought. Whenever pebbles occur mixed in clay soils, they are a hindrance to mole drainage, but they break up the soil and make cultivation easier. As a general rule clay soils have a pH of 5·5–6·5. This level of acidity is satisfactory for growing most farm crops—see Table 2.1—and it is unusual to consider applying lime or chalk to correct acidity, but these alkaline substances may be used, occasionally, to improve the texture, and so help drainage of a very "sticky" clay soil. As a rule clay soils contain adequate supplies of potash available for the growth of all crops; but they should receive phosphates for many crops, and especially for the legumes such as beans on the arable land, or for the clovers on the grassland or the leys.

Sands and Gravels

The simplest way of explaining the difference between sands and gravels is that both consist of coarse particles; but with gravels there may be many small stones, whereas sands contain no stones. The stones are an advantage since they anchor the soil and very definitely reduce wind erosion—sandy soils that contain no stones may suffer very badly from blowing. Both soils are so very abrasive to implements that, when ploughing, shares must be changed frequently.

The worst sands and gravels are only suitable for trees—conifers—but the slightly better soils will grow rye, oats, potatoes, carrots, barley and peas. The tolerance of the various crops to acidity is clearly shown in

Table 2.1, though rye will grow on soils with a pH of 4·5. It is the acid-tolerant crops, oats and potatoes, that will be found on the most acid light sandy soils. This does not mean they are the only crops which can be cultivated, but if other crops are to be grown then the acidity must be corrected. Thus on acid sandy soil, lucerne has been grown quite successfully after lime or chalk has been applied. It must also be stressed that once the acidity has been corrected it may be necessary to repeat the dressing at about 5-year intervals, for the neutralising material is washed down in a very few years and so may be too deep for the roots, especially the seedlings of some crops, to reach it. The current use of artificial manures results in the available calcium being used up more rapidly and so increases the need to add fresh supplies of lime. Experimental work has shown that on these soils, for both roots and cereals, liberal manuring with all nutrients is essential for maximum returns. Owing to the free-draining nature of sands and gravels it is essential to apply nitrogenous manures for all crops, and to assume that it is unwise to apply any artificial manures liberally with the hope that any surplus will be left for the subsequent crop—in effect, in winter, surpluses will usually be washed out of the soil and be lost. It is not surprising that under these circumstances the sands and gravels are known as "hungry soils"; they give good responses to artificial manures when applied at frequent intervals.

Loams

Where there is a mixture of clays and sands, giving loams, then a big variety of crops may be grown. It is not necessary to enlarge upon these soils since the cropping is relatively straightfoward.

Chalks

Chalks may be of pure chalk or chalk and flints; the latter assist with cultivations when under the plough. Because chalk soils are free-draining, they may be ploughed and cultivated at many times during the year, though they may be very "sticky" just after rain, and under such circumstances they should be left till they have dried before cultivations are done. Since they are free-draining soils they, too, are also known as "hungry soils", because they give such good responses to artificial manures when applied at frequent intervals.

During the past 50 years the greatest changes have occurred in the cropping of chalk soils. Originally the worst fields were rabbit runs and used for sheep walks. After the rabbits were removed and the land was improved by manuring, much was sown down to grow grass for dairying, for breeding pigs or for laying hens. Since these leys were often grazed at some time during the period they were kept down, and in many cases were grazed more or less continuously, it is apparent that the humus content was increased and so the soils were much improved. After they had been improved, the leys were ploughed up and cropped with a variety of arable crops, but frequently the grassland was retained for dairy cows and for young stock.

Peats

The peat soils are very high in organic matter and they are naturally sour; the worst are given over to rough grazing. To bring such land into cultivation would be so expensive with liming and the liberal use of artificial manures, that it is seldom reclaimed; it may provide a good return for letting for game shooting. Peat soil that is not too sour, however, is reclaimed and used for growing oats, potatoes and grass. It has been found in some instances that it is profitable to sell off the peat for fuel, or for litter; the remaining land can then be brought into arable cropping fairly easily after some of the peat has been removed.

Fens

The other soil type that is high in organic matter is the fen soil. There is a large area of this soil in East Anglia, but in addition, in many river valleys and marshy areas, there are small pockets of fen soils. Since these soils are low lying (often below sea level) they are drained by expensive drainage schemes that include digging main water courses (drains), maintaining river banks, and installing expensive pumps to lift the water from the drains into the rivers. To meet these charges farmers are levied with drainage rates annually that may be several pounds per acre—like paying an additional rent. In times of drought, water may be run from the river into the drains for irrigation purposes. Such soils are valuable because of their natural fertility, and provided the drainage is properly done, these soils grow a large range of crops, especially root crops; of the "root" crops celery is very important as it can be very remunerative. The value of fen soil depends on what is beneath it. The best fen soil has clay beneath; the worst, sand and gravel. The first main difference arises from the fact that if clay is beneath then the land will never suffer severely from drought, but if there is sand and gravel beneath then drought is a real risk. The second difference lies in the fact that if the fen soil contains a good proportion of clay it will not suffer from wind erosion, whereas a fen over sand may suffer so badly that annually in the spring crop after crop may be blown away, and the only real remedy is to add clay to the soil. This is an expensive operation if the clay must be transported any appreciable distance and in addition it may have to be applied at the rate of several hundred tons per acre.

It is sometimes said that fen soils are "kind" soils for there is very little wear upon the equipment used for cultivations. A ploughshare may last all the season. On the other hand, in springtime the soil is light and fine (when dry) and when doing cultivations it blows around the tractor, thereby making it essential to change the air filter at relatively frequent intervals.

Mountain Rocks

It has been mentioned that there are, in the mountainous areas of the United Kingdom, vast stretches of countryside which are just rocky outcrops with little or no soil at all. If this is the case there is no possibility of the land being used for anything other than for amenities, such as rock climbing or, when of suitable slope and covered with snow, for ski-ing. If the land carries a small covering of soil then it may be possible to grow trees. If the soil is still a little more abundant, then grass will grow and provide grazing for breeding sheep. In some instances such land may only carry 1 sheep to 3, 4 or more acres, but if it is improvable then it may carry 1 sheep per acre. On the best mountain grazing, single suckling beef breeding herds may be found. These mountainous soils play a far greater part in British agriculture than is sometimes believed, for they cover a large area, and they are the breeding ground for the majority of breeding ewes that are found on most lowland soil types. They also provide large numbers of wether lambs that are subsequently fattened on both lower arable and grassland.

The British Government gives various subsidies to encourage breeding of sheep and beef cattle on the hills and to assist in providing the essential winter food for the breeding stock, so that these mountain and hill areas contribute their share to the national larder.

While mentioning mountains and hills it should be pointed out that if there are very steep slopes, where the soil is quite good, it may be impossible to plough the land to grow crops. In other words, topography may seriously affect the way in which the land is used. Steep slopes may be unploughable, but if there is reasonably good soil then the resulting grass will make satisfactory grazing. Where a level area is found it may be sprayed to kill the "turf", lime and artificial manures are applied, and a seeds ley mixture may be sown into the turf. This has proved to be an effective way of improving an area. Though it is now agreed that fencing, at least in the initial stages, is an advantage, the benefits of treading and manuring these small areas may spread on to the adjacent land around and so extend the benefit from improvement. However, as a general rule, where steep slopes are found the good soil has been washed down into the valleys and this better soil has been used for growing arable crops for home consumption, whereas on the slopes only grass will grow.

Table 2.1
Crops and pH tolerance

Crop	pH tolerance
Rye	4·5–5·0
Potatoes	5·0–7·5
Oats	5·0–6·5
Winter wheat	5·5–7
Continuous cutting or paddock grazing	6·0–7·5
Mixed leys	6·0–7·5
Lucerne	6·5–7·6
Beans	6·0–7·5
Swedes	6·0–7·5
Kale	6·0–7·6
Feeding or malting barley	6·1–7·5
Sugar-beet	6·1–7·8

Mixing Soils

In some cases soils are mixed artificially in order to improve them. Very frequently lime or chalk is added to the sands and gravels to correct acidity to grow certain crops (see Table 2.1). In general the quantity applied is small and will have to be repeated at intervals of from 4–10 years depending upon the acidity and upon the cropping policy. On fen soils clay may be added once in every 30 years to prevent blowing. So important is this claying of fen soils that a government grant is given to assist in this very expensive treatment.

Most soils are modified to a lesser degree by the addition of organic manures, such as farmyard manure, green manuring and by ploughing in crop remains, e.g. turf from leys. While some farmers burn the straw from their cereals, others spread that straw and plough it in, sometimes with artificial manures added (nitrogenous) to assist with the rotting.

Manuring

In the case of the growing of some fruits, flowers and vegetables, organic manures may be used. Some are quite expensive, for example, shoddy, meat and bone meal and spent hops, and these should only be used on crops of a high cash value. When such organic manures are used

Table 2.2

Average requirements of crops for nitrogen, phosphorus and potassium

(These general figures* would need adjustment for soil type, climate, region, use of farmyard manure, and previous crop)

Crop	Average units of of N, P and K applied		
	N	P	K
Potatoes	129	140	190
Oats	38	29	26
Winter wheat	64	39	35
Continuous cutting or paddock grazing	207	55	40
Mixed leys	71	55	39
Lucerne	0	55	39
Beans	0	40	80
Swedes	30	40	40
Kale	50	40	40
Feeding barley	53	32	33
Sugar beet	135	130	180

Notes Many factors are involved in balanced fertiliser application for any crop The above figures should therefore be treated as a broad guide only.

* 1966 figures for the United Kingdom.

for a period of years on a field they can materially change the nature of the soil. Finally, most soils require nutrients to grow certain crops, and these nutrients in the form of artificial manures must be given freely in order to obtain maximum results. The actual proportion of these nutrients given depends on the soil, the climate, the crop, and the manuring of previous crops in the rotation. This is clearly shown in Table 2.2, where the variations in usage according to the crop are considerable. The ratios of nutrients given will not remain constant for they will be linked to factors like the various crop needs and the soils on which the crops are grown.

Greater use of nutrients and a change in manuring policy may be profitable. Manuring policy should be planned to meet, as far as possible, the requirements of the crops grown, bearing in mind existing levels of fertility in the soil and the type of soil and climate involved.

CLIMATE

Frosts and snow

Soil is not everything, for climate will have a major effect on crops and cropping. As a general rule the climate of the United Kingdom may be classed as temperate, but it is true that many variations are to be found; for example, while snow is rare in the south-west in winter-time, it is more common in the north (in the Aviemore district of Scotland, ski-slopes are being developed for use for half the year; their height puts them outside the farming range). In many arable districts frosts are common in mid-winter, but it is rare that these frosts are sufficiently severe to damage crops. It is the spring frosts, and the risk of late frosts, that may change the cropping policy. For example, it would be unwise to grow fruit and some flowers in an area where there is a risk of late spring frosts, which might damage the blossom or the newly-set fruit. Experience shows that certain fields with southern slopes are less prone to frosts whereas, in the same locality, the northern slopes of a hill are liable to frost attack. Sometimes the early autumn frosts may damage maize and celery; the ordinary farm crops rarely suffer except with very severe frosts when the sugar-beet crop is either still in the ground (and then lifting may be very difficult, or impossible) or when the beet has been lifted and stored in large heaps. Good coverings of straw will usually give adequate protection, but some farmers make no preparations and take the risk—and sometimes they have to pay for their neglect, for the factories may refuse to accept frozen beet because they find the sugar cannot be completely extracted.

Rainfall and irrigation

Usually there is sufficient annual rainfall for the growing of crops in the whole of the United Kingdom, but it is not uncommon, on the light soils, to find in some years that there are early summer droughts so serious as to affect crops when they are growing. Such droughts can reduce yields of cereals, grass, and such vegetables as peas and potatoes. On average it pays in the areas east of a line drawn from the Isle of Wight to the Wash to use irrigation when required on the more valuable crops. It is not that the rainfall is too light but that it falls at the wrong time. To ensure that water is available for irrigation, some farmers buy it from River Authorities, while other farmers have built reservoirs to store winter rains for use in the spring and summer. This is wise planning and organisation. Other areas may also benefit at times from the use of irrigation, but farmers must decide whether, on average, they should provide it or not. Shortage of summer rains may also lead to a serious shortage of grass for dairy cows. It is not uncommon, too, to find that severe droughts in the late summer make it essential to feed cows on winter rations prematurely. This may extend the winter feeding period by 2 or 3 months under the worst conditions.

Temperature

In many areas of the United Kingdom the climate is sufficiently mild for grass to grow for 8 or 9 months of the year, and in such cases the stock may remain outside all the year round, provided the land is not too heavy and will not suffer from stock poaching the surface. It may be necessary to supplement the grass in the severe weather in winter by hand-feeding. This may be only for a few days, but if it is for many weeks then the stock is better taken indoors, if buildings are available. In the south of England in sheltered areas even dairy cows may remain out for the whole year, while in the north, and in the exposed areas, they may be in for as much as 6 months.

With some of the modern large dairy units experience has proved that it may be advantageous to keep the herd indoors all the year round, not because of the climate, but to prevent the cattle from damaging the grassland and so reducing the yield from the grass. Grass

INTRODUCTION TO FARMING SYSTEMS 27

is cut, and carted to the cows indoors—a system which is known as "zero grazing", and which is practised by some farmers for the whole year.

For ordinary farming it is not possible to raise the air temperature in an area, but farmers resort to irrigation of fruit blossom to reduce frost damage. The irrigation of water meadows with warm spring water has given early growth of grass in the spring. The only other way in which land can be warmed is by draining any that is wet and waterlogged; land that needs draining produces late crops.

LABOUR

Falling labour supply

Annual statistics show that the labour force engaged in British agriculture is falling; the area of land that is farmed is being maintained, and there is evidence that production is becoming even more intensive. This means the remaining labour force is being used more efficiently than formerly. This is achieved by the use of machinery. Larger tractors and bigger equipment enable men to achieve a bigger output per workday. This has been the policy for the last 20–30 years. Added to the falling labour force has been the fact that men have worked for a shorter number of hours per week and this has made it even more important to achieve a bigger output for each hour worked. The use of the larger equipment has been made more efficient by enlarging fields by the removal of hedges and boundaries; there has also been the policy to amalgamate farms. This has involved farmers in large expenditure, and to meet this the Government has provided grants for the removal of hedges and for the amalgamation of fields and farms; but this has not been sufficient in some cases because of the high cost involved of equipping farms with larger equipment. To meet this farmers have sometimes formed co-operatives or syndicates to share equipment, which has worked well with slurry handling equipment that is used throughout the year, but less well with machinery such as combine harvesters which work only for short periods annually. There has also been the development of the contracting business by which engineering firms have provided the bigger equipment required on the farm to do such work as drainage, levelling, bulldozing, ploughing and combining. During the last few years some firms selling artificial manures have provided their customers with a service to apply the manures on a contract basis; there is evidence that this system is increasing in popularity for the ordinary artificial manures. It is essential with liquid fertilisers where, in order to achieve the best results, the liquid should be injected into the soil; for this special equipment must be used.

Labour has become expensive and this has accounted for certain crops that can be easily mechanised becoming popular, e.g. cereals; whereas the crops that cannot be so easily mechanised, e.g. mangolds and swedes, have become less popular. It is possible to pick (under certain circumstances) blackcurrants and raspberries by mechanical means, but it is not possible to pick cherries, so the acreage under cherries in the U.K. is declining for this very reason. The only way, so far, to pick strawberries is to do so by hand; one way in which the labour problem has been overcome has been to allow customers to pick their own fruit and to pay for it at a reduced price per pound when leaving the field in which the fruit has been grown. It would seem likely that this method of picking fruit, where mechanisation is impossible, may increase.

Gang labour

Farmers must study the available labour situation before they can embark upon labour-consuming enterprises. If the labour is not available from the households of the farm workers it may be necessary to explore the position in the nearest village. For such enterprises as fruit picking, or the seasonal vegetable and flower picking and marketing, farmers may have to transport workers to and from their homes to the farm, otherwise there may be no labour. It is quite apparent that in such cases the labour can be a very serious limiting factor. In some areas of Great Britain, particularly in the vegetable growing areas of East Anglia, men (gangers) will collect a team of men or women or a mixed team to move from farm to farm to harvest certain crops, e.g. potatoes, carrots and brussels sprouts. These gangs will harvest and grade the produce for market and since they specialise in this work they become very efficient at it. Such gangs usually demand to be paid piece-work for what they do. The ganger usually makes all the arrangements with the farmer and produces the gang when required. These gangs travel quite big distances—10 and 15 miles from their homes are quite common.

Relief workers

It should also be pointed out that in a normal season, except at harvest, or when work has been hindered by bad weather, it is possible to work for 5 days a week with crop production, but with livestock this is not practicable. Those keeping livestock must in general organise shift-working of labour at the weekends to provide relief workers, otherwise farmers will be short of workers. In some instances the stockmen have agreed weekends off in rotation, but sometimes one finds a very keen cowman who prefers to work every day but instead of having a weekend off he prefers to do less each day (for example to have time off between milkings). He may receive a bonus according to the milk production of his cows and he may prefer that there should be no change of milker since that may result in a loss of milk production. Once milk is lost from a cow for even one milking, it is usually difficult to regain the yield.

Housing farm workers

During the last 30 years there has been a change regarding the housing of farm workers in Great Britain. Formerly they lived on the farm in what were known as "tied cottages" for which a nominal rent, or even no rent was charged. The staff engaged with crop production need not now live on the farm since horses are no longer used, and consequently there is a growing tendency for such workers to live in villages and to travel to the farms daily to work, sometimes on bicycles and frequently in cars. Town dwelling is a great convenience for children going to school and for the wives to do their shopping. Gang labour will also live in the town or village, and because of their skills will move from farm to farm to do the work required. It is still generally accepted that the stockman should live on the farm in order to see his stock last thing at night, and to be available for sick stock and for calving cows, lambing ewes, or for farrowing sows. This means that in some cases the stockmen become the only men who live on the farms. One wonders how long this will persist, for this may act as a further deterrent to young men becoming stockmen if they see they are "tied" to the farm.

Wages

In order to meet the labour problem, British farmers must pay good workers at wages considerably above the minimum wage as laid down by the Wages Board. This may be as a weekly wage according to the number of hours worked, with due allowance for overtime, but in some districts it is worked upon a piece-work basis which provides an incentive to increase output and which will enable keen workers to earn 25–50 per cent above the ordinary daily wage. In some districts most of the farm-work with crops, fruit, vegetables and flowers is done on this basis; but in other districts, where piece-work is not usually worked, it is viewed by the workers with suspicion and they will not usually try it. Piece-work results in greater output for which higher wages are paid, and this helps the farmer to get the work done more quickly while it also pays the worker extra wages. It should be an advantage to both partners, provided the work is being done properly. In crop production, weather is often a limiting factor and speed may be of great importance.

Output per man

With the increased cost of labour, British farmers find it essential for each stockman to look after more stock than formerly. This has been most marked in the case of dairy cows, where numbers looked after by one man have increased 8 or 10 times in the last 30–40 years. This means a stockman may have to be more observant and more skilled than his predecessor. In order to be skilled in this way a stockman may need a special education so that he is a fully qualified technician. Much the same may be said of the skilled tractor driver, who is required to operate, maintain and service big and expensive equipment, sometimes worth several thousands of pounds. Men must be fully trained and paid for these special skills.

To meet high labour costs, more and more farmers become specialised, and both pig and poultry units have moved from the farm to what may be described as "factory" units. It is found that by keeping stock under factory conditions (on concrete) one man can deal with many animals and so labour needs can be reduced. So far factory farming has been applied only to pigs and poultry. Already it is operating to a limited extent with the fattening of cattle, both for veal and beef production, but there are signs that it might also apply to dairy farming where zero grazing is practised. In a few instances the same general principles have been applied in the winter time to breeding ewes kept indoors from the stronger soils and hill grazings. So the suggestion may be made that factory farming may be appearing amongst all classes of livestock in this country—a main object being to reduce labour costs.

ECONOMICS

The market

It has been pointed out that any farming system is dependent upon the supply of labour, either on the farm itself, or near enough for the workers to come to the farm of their own accord or be transported by the farmer. There is another aspect that will influence cropping, namely, the possibility of selling the produce in an appropriate market.

Some farm produce may be transported thousands of miles from the point of production, e.g. butter, cheese and dried milk are various long-keeping milk products. To this list may be added frozen meats, cured meats (bacon) and all the various tinned meats. Cereals, if well-dried, will travel from the Antipodes. With modern gas and cold storage, fruits can travel from all ends of the earth to the United Kingdom.

It may seem that local demands need not be considered, but fresh fruits and vegetables will travel at most only a few hundred miles. For the newest process of quick freezing of such vegetables as peas, they must be grown within 15 or 20 miles of the factory in which they are to be processed. Experience shows that if the peas are to retain a good colour they must be processed within minutes of being threshed in the field in which they have been grown. This freezing process is fairly new, and is also proving successful with other crops such as brussels sprouts, celery and runner beans. If one is anxious to grow crops for the new quick-freeze trade one must be sure that a factory is near enough to make it possible to obtain a contract, which ensures the right varieties will be grown with assured sale if satisfactory produce is obtained.

Price of land

There is yet another side to this question of economics. During the past twenty years, the sale price of land in the United Kingdom has been rising fast, though since the early part of 1969 there was a slight recession lasting for one year. When a high price has been paid for land either to rent or to purchase (and especially if the purchase money has been borrowed at a high rate of interest), it is essential to turn to such enterprises as will produce the return to enable the borrower not only to pay interest (possibly at a rate as high as 10 per cent or more) but also to repay gradually the original capital that has been borrowed. This means that mainly crops which give a high return can be grown, such as the root crops, vegetables, fruit and flowers and crops for seed production, while there may be very little opportunity of obtaining a high return from keeping stock in a traditional way on grassland, or leys. To obtain the maximum return from stock it will be essential to keep them intensively to achieve high output per acre. With the high price of land no extensive system of keeping livestock can be considered; on the other hand, in a locality where most land is high-priced, some may be at a lower price and then an extensive system may be practised.

PERSONAL INTEREST

There is one overriding factor that must influence the system of farming that is adopted, namely, it must enable the farmer to carry out some system, or to keep the stock or to grow the crops, in which he is really interested; for unless there is real interest there will not be success. Thus any farmer is well advised to try to find the farm that will enable him to carry out the farming to which he is attracted. In these days when margins are small, enterprises must be very efficiently run and this is only possible if attention is paid to essential details. This is more easily achieved by a farmer who is intensely interested in the particular enterprise, or enterprises. If there is a choice depending upon the land, climate and labour, it must be where the real interest lies.

CAPITAL

All farming needs working capital in addition to the initial outlay or the purchase of land, equipment and stock. Many farmers, especially when they are starting farming, are short of capital and so cannot farm in the way in which they would choose, but they must farm in such a way that they can build up both working capital and also other capital for purchases. In the past it has been recognised that one of the best entries into farming in Great Britain was by way of dairy farming, since that gives a monthly cheque for milk sales and this provides the means of paying for the labour and the concentrates (paid monthly). This gives a quick and steady turnover. The same farmer will be well advised (except for diseases) to purchase down-calving cows or heifers in order to build up his herd rather than to retain heifer calves that will not in general give any return until they have calved at about 2 years of age. Similarly he will not be able to start with a pedigree herd, but he can grade up from non-pedigree to pedigree within 4 generations, or later gradually to buy pedigree cows or heifers. A much quicker turnover can be obtained if weaner pigs are purchased and fattened than will be the case if the weaners are home-bred. The latter plan will add a further 6 months before the weaners are ready for fattening. The weaners give the quicker turnover and so should be kept initially even if breeding is the ultimate goal.

When starting farming, when capital is usually in short supply, some farmers avoid buying expensive new equipment either by purchasing essentials secondhand, or by hiring contractors to do certain jobs on the farm; and they also co-operate with their neighbours to form syndicates for buying and sharing equipment. Another way in which limited resources are husbanded is by either borrowing from a bank or by arranging with merchants to pay for goods (e.g. seeds and fertilisers) after produce has been sold. This means that sometimes credit is needed for 6–12 months, but of course this leads to loss of discounts. For a time the farming system must be geared to the resources available, and not always to the system that may prove most remunerative in the long run.

SYSTEMS AND GOVERNMENT POLICY

The farming policy adopted on any soil must be dependent on the returns obtainable for the produce from that soil. In the United Kingdom, this is demonstrated by the changes in husbandry that have appeared in the last 40 years—the old system of keeping sheep on arable land has gone, and in its place different breeds of sheep have been kept on leys, instead of on root crops as before. Farmers, by economic pressures, have been forced to keep stock in bigger units so that more stock are looked after by each man; to achieve this it has been essential to modernise farm buildings.

On the arable land the combine harvester has replaced the reaper/binder and threshing machine. In other words, systems that cannot be sound financially disappear, but the whole profitability of farming is very much dependent on Government policy. British Government policy at present (1972) is to discuss and arrange with the National

Farmers' Union (who represent the farming community at the February Price Review) the prices of the principal agricultural products for the succeeding year. By price fixing incentives the Government emphasises what farmers should produce. It is also Government policy to decide which foods should be imported, the countries from which they should come, and quantities to be imported from each country. In this way the British Government is able to foster trade and formulate the agricultural policy of this country. In general, farmers consider the Price Review system has worked fairly well but they feel that to have an annual change to put emphasis upon the crops that should be grown, and the stock that should be produced, does not provide a sufficiently long-term policy. Consequently farmers sometimes lack confidence to pursue policies that demand considerable expenditure of capital. Farmers would be more confident if a programme for say 3 or 5 years were announced.

The two basic points that rule Government decisions are:

(1) The general public must live as cheaply as possible and the Government must decide whether food should be produced in the United Kingdom or imported,
and
(2) the Government must decide whether British farming is to be maintained or not.

Sometimes foods are imported to the United Kingdom from other countries at very low prices because they are surplus to local requirements; sometimes such foods receive subsidies from the foreign Government, and consequently they may be "dumped" in the United Kingdom at ridiculously low prices. British farmers object to this policy and have asked the British Government to prevent "dumping". If farming is to continue on a sound basis, farmers must have confidence and be assured they will have adequate returns for the efforts they put into farming.

In subsequent chapters the influence of soils on farming systems will be explained in greater detail; but in all cases success depends upon the individual in charge, his ability as husbandman with both crops and stock, his ability to handle labour and his business acumen to buy and sell on the most favourable markets.

Chapter 3

ALL-ARABLE SYSTEMS

The climate of the British Isles allows a wide range of arable crops to be grown. Some of the main ones are cereals; root crops such as potatoes, carrots and sugar-beet; peas; hops; and brussels sprouts.

Provided drainage is satisfactory, the main limiting factors (apart from climate and topography) are soil type, availability of labour and in some cases availability of capital for machinery and buildings. For example cereals will grow on all ploughable soils and have the lowest labour requirement. Whether an arable farm includes some grassland or none at all depends among other things on the type of soil. Cash "root" crops such as potatoes and sugar-beet will only grow satisfactorily on deep and preferably rich soils; these crops have much higher labour and capital requirements.

The actual amount of labour required is closely related to:

the degree of mechanisation which is possible;
the availability of reliable casual labour for rush periods;
the layout of the farm;
the size and shape of the fields.

Economic returns are also very important in deciding which crops might be grown; this question is related to the size of the farm and its management. These aspects of cropping systems are dealt with in the chapters on management and mechanisation. Quotas and contract acreages are important for some crops, e.g. potatoes, sugar-beet, vining peas and hops.

The main arable crops grown, their acreages and general distribution are shown in Figure 3.1.

Cash cropping, without grazing livestock, has traditionally been practised mainly in the eastern parts of Great Britain because of the normally sunnier and drier weather conditions in these areas which make harvesting easier. On some all-arable farms, winter fattening of beef cattle provides farmyard manure for the root crops; pigs may also be kept to increase the profitability of the farm as well as providing manure.

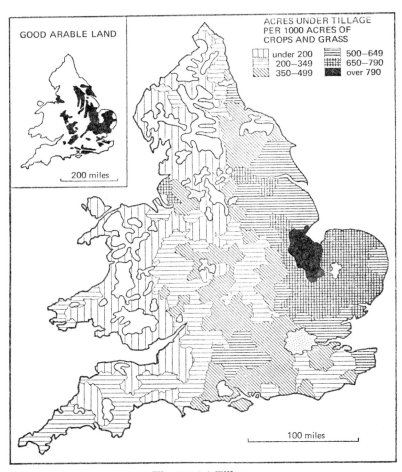

Fig. 3.1 (a) Tillage

ALL-ARABLE SYSTEMS

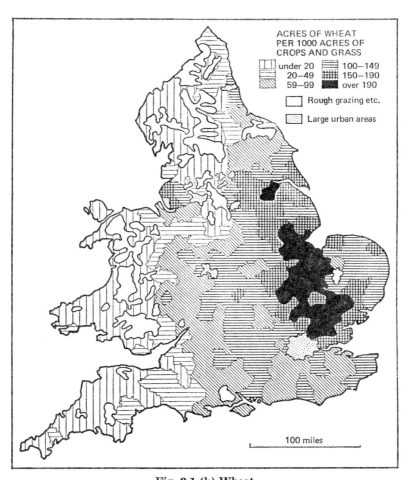

Fig. 3.1 (b) Wheat

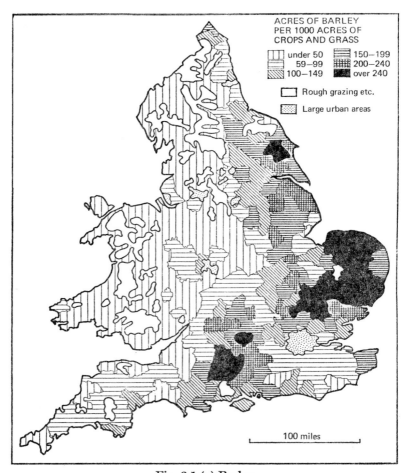

Fig. 3.1 (c) Barley

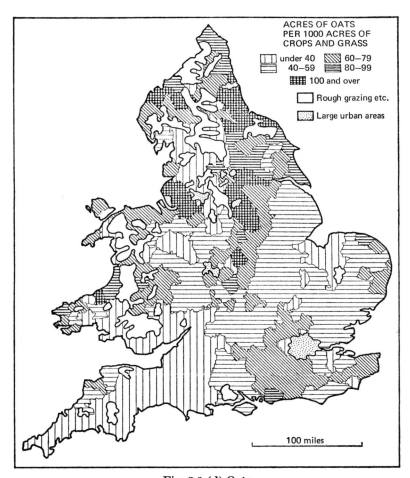

Fig. 3.1 (d) Oats

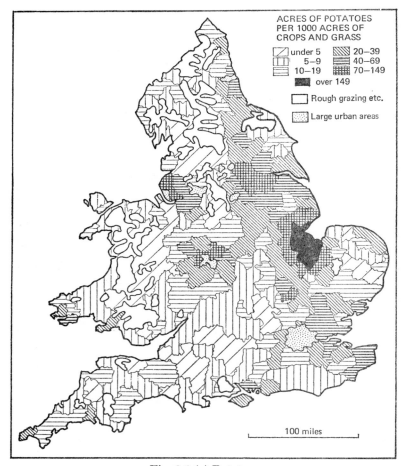

Fig. 3.1 (e) Potatoes

ALL-ARABLE SYSTEMS

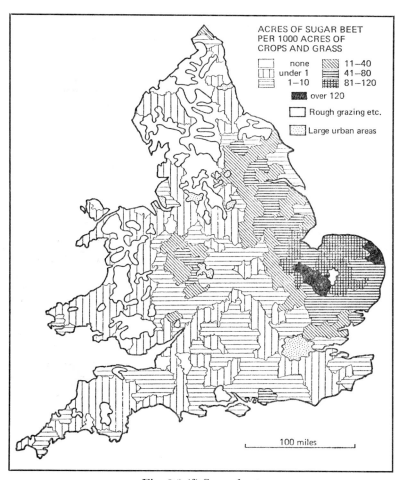

Fig. 3.1 (f) Sugar-beet

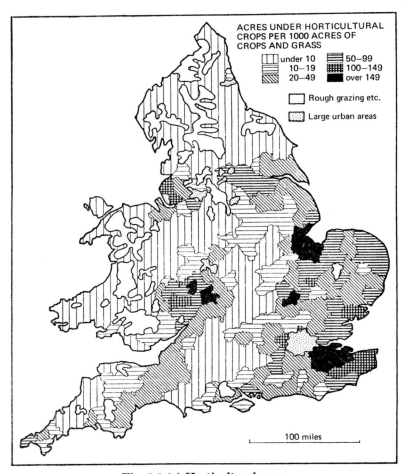

Fig. 3.1 (g) Horticultural crops

ALL-ARABLE SYSTEMS

In recent years, all-arable systems have become more widespread. They are now found in most counties and on a wide range of soil types. In some cases, only part of a farm may be continuously in arable crops. The main reasons for the increase in arable cropping are:

shortage and high cost of skilled labour for livestock enterprises;
the possibility of a 5-day working week for most of the year, compared with a 7-day week which is essential where livestock are kept;
the lower profitability of traditional sheep and beef enterprises;
shortage of capital to finance livestock enterprises;
greatly improved mechanisation, particularly the introduction of combine harvesters and drying facilities for cereals;
the introduction of a wide range of herbicides to control many of the troublesome weeds in cereals and root crops; this saves labour and reduces cultivations to the minimum required for growing the crops.

SIMPLE OR EXTENSIVE ALL-ARABLE SYSTEMS

Extensive systems such as continuous cereal production, or slightly modified systems including occasional break crops, are now found in nearly all areas where arable cropping is possible, mainly spring barley on the loams and lighter types of soil, and winter wheat on the heavier soils. Where it is not desired to have continuous cereals on the whole farm, then part may be cropped in this way for reasons such as remoteness from the buildings, soil too shallow for root crops or lack of water for stock.

SOIL-BORNE PESTS AND DISEASES

An all-cereal system has relatively low labour and capital requirements, but is risky because of losses, which can be serious, owing to build-up of soil-borne pests and diseases. The most troublesome pest is cereal cyst (root) eelworm (*Heterodera avenae*), and the most troublesome fungoid diseases are take-all (*Ophiobolus graminis*) and eyespot

(*Cercosporella herpotrichoides*); foot-rots (*Fusarium*) and sharp eyespot (*Corticium solani* or *Rhizoctonia solani*) may also cause trouble.

Take-all and cyst eelworms attack the roots of the cereal whereas eyespot attacks the stems (straw) a few inches above the soil. Their general effect is to interfere with the normal absorption or movement of water and uptake of nutrients, with serious effects on the grain-producing capacity of the plant. In addition, they cause premature ripening of the infected plants so that their effects are easily spotted as the season advances.

Take-all

All barleys and wheats are susceptible to take-all, though barley varieties are not so severely affected. Oats are attacked by a sub-species of the fungus that is found predominantly in the West and North of Great Britain, so in the central and eastern arable areas this crop is suggested frequently as a break-crop; though it must be realised that it is very susceptible to cyst (root) eelworm which might well be present. Oats are also very susceptible to stem eelworms (*Ditylenchus dipsaci*) against which resistant varieties are available.

Take-all is a facultative parasite which can, to a limited degree, survive saprophytically on decaying matter when a suitable host plant is not available. Thus the use of spring barleys has tended to reduce the impact of the fungus in continuous corn growing, because spring barleys are less susceptible and occupy the ground for only about half the year. For the rest of the time the fungus must grow saprophytically on decaying trash. It is therefore sound practice to plough or cultivate unburnt stubbles early, and to bury plants which establish themselves from shed grain—so promoting quick decay.

Winter corn crops can act as a "bridge" for take-all and aggravate the situation. Unfortunately, couch grasses (*Agropyrons repens* and *Agrostis gigantea*) can increase under all-corn conditions and these species together with creeping bent (*Agrostis stolonifera*) can act as hosts to take-all; hence the importance of controlling these weeds.

The Chamberlain family have practised continuous corn growing in Oxfordshire for a very long time, and have developed a system of undersowing corn with yellow trefoil and/or Italian ryegrass, which are ploughed in as green manure after harvest. This appears to give some control over the build-up of take-all. It seems that this practice encourages non-pathogenic soil saprophytes that break down sugar-rich plant tissues and these fungi have an antagonistic effect on take-all during its saprophytic stage.

Take-all spreads rapidly and is usually most frequent and severe in alkaline and coarse-textured soils, especially those overlying chalk or well supplied with lime.

In fertile soils new roots can develop rapidly to replace those killed by the fungus, and in heavy soils the fungus does not spread so rapidly. On badly drained fields, the effects of take-all are often more severe because the waterlogged conditions reduce the vigour of the wheat or barley crop and restrict the development of new roots to replace those damaged by the fungus. Where soil fertility is low, the smaller root systems are easily destroyed.

An interesting recent development is the slit-seeding of cereals into unploughed, paraquat-treated, stubbles. Compared with continuous corn cropping there is a slower build-up of take-all, not from the fungicidal properties of paraquat, but probably as a result of the firmer seed bed limiting the spread of take-all hyphae.

Studies have shown that, as successive cereal crops are grown on a field, take-all usually increases to a peak and then declines. The reason for this decline is not clear (the parasitic behaviour of the fungus may be inhibited in some way), but the recession is to a level that still seriously reduces the corn yields when compared with those obtained in normal rotations, perhaps by as much as 15 cwt per acre. As the prevalence of take-all levels out, so therefore does grain yield, but there can be marked seasonal fluctuations; good and bad take-all years occur. It is possible to have an almost complete crop failure in some seasons, (usually wet seasons), and little harmful effect in other seasons. It would be valuable if the factors causing these variations could be fully determined.

It might therefore be sensible to embark on continuous corn growing with spring barleys, which can weather the rising and peak years of the fungus, and then when the peak has passed, to introduce the occasional winter wheat crop as a more remunerative cereal. It would also be useful if the wheat variety used was resistant to eyespot, so acting as a break to this fungus too. If continuous wheat is being grown there is no advantage and every disadvantage in delaying sowing; the stubbles should be ploughed and new crop sown before mid-October if possible.

If, under these conditions of cereal growing, a non-cereal break crop is introduced, its effects are not likely to be as simple as merely reducing the take-all. If the situation is examined more closely it is seen that the break-crop reduces take-all immediately and the cereal yield may rise by 15 cwt per acre. However, take-all now builds up very rapidly and again rises to a peak whereas it had more or less settled at a lower level of infection prior to the break. Thus the second and third cereal crops after a break may have poorer yields than the pre-break yield. The situation can be summarised as follows:

Crop year	Yield (cwt/acre)	Effect (cwt/acre)
Pre-break yield	approx 25	—
1st crop after break	,, 40	+15
2nd crop after break	,, 20	−5
3rd crop after break	,, 20	−5
4th crop after break	,, 25	—

The total benefit of the break is an extra 5 cwt per acre during a 3-year period. In addition, a year's yield of corn has been lost, resulting in a loss of 20 cwt over the whole sequence. The break-crop must therefore be one that is very profitable or has some other outstanding advantage, such as allowing better weed control.

Eyespot

Eyespot attacks all cereals (but oats least of all) and is usually most damaging on winter wheats, especially if the autumn is cool and wet. Early sowing, and good seed-beds to promote vigorous root development, will help such crops; but most important is the growing of those wheat varieties that are least susceptible to this disease. The use of Cappelle-Desprez for continuous winter wheat production has been as successful as it has because eyespot does not build up seriously in it. In fact, when spring barleys are being grown, such a wheat variety may be used as a "break" against eyespot.

However, eyespot is often severe in its effects, especially on heavy land. It is spread by spores that are splashed usually from infected litter and plants. Lodging of the crop often results in the straws being laid in all directions, when the poor quality grain is easily picked off by birds. Premature ripening (or "whiteheads") is also found, as with take-all; and indeed, it is usual for both diseases to occur together in fields where successive corn crops are grown.

Sharp eyespot (*Rhizoctonia solani*) is very frequently encountered. It is often confused with eyespot (*C. herpotrichoides*), though it can be distinguished by the generally clear centres to the stem lesions, which also tend to spread higher up the straws. Sharp eyespot can be present to a greater degree than eyespot; winter wheat with 10–20 per cent of infected straws in continuous cornfields is common.

As with take-all, early ploughing, the reduction of grass weeds, good seed beds, and not sowing winter wheats too late nor spring barleys too early, help to keep infection down; though with winter wheat, less susceptible varieties are very important.

A break in cereal crops reduces the incidence of eyespot, but it needs to be a break lasting several years. Eyespot has a longer period of survival on buried plants and trash than other facultative cereal diseases, and therefore short breaks are ineffective.

Cereal cyst eelworm

Cereal cyst eelworm is a widely distributed pest, and can be particularly troublesome on light soil. It can seriously depress yield as it builds up in continuous corn production. This can be demonstrated by the improved yields obtained by sterilising the soil with formalin or other soil fumigants. The eelworm persists in the soil as egg-containing cysts, which hatch in the spring and invade the cereal roots. Barley is more tolerant than wheat; oats are very susceptible, and so may give very poor yields when introduced as a break in a series of barley crops.

Eelworm infestation causes a characteristic patchiness in the crop from quite an early stage, and during late May patches of stunted plants can be obvious. Earlier this may be manifest in the yellowing of leaves. The effect of the root invasion is that it stops root extension, causes branching and leads to a shallow, matted root system so that if the plants are pulled up they bring a ball of soil with them. The affected plants form smaller ears and shorter stems than uninfested ones, and in June–July swollen immobile females develop in the roots. At first these are very small; later they are seen as white glistening protrusions, that fill with eggs, harden and form cysts which are brown and become detached from the plant, resting in the soil. Chemical controls are not practicable.

Suitable resistant varieties are now available, e.g. Sabarlis, and further breeding for them is being actively pursued at plant breeding stations. The situation is complicated by the known occurrence of different biological races of the pest, some of which can overcome certain types of resistance. The use of good fertiliser dressings on barley crops— particularly the use of high rates of nitrogen—can help the crop survive quite large soil-borne populations and give satisfactory yields, but even so there can be a high percentage of poorly filled grains. In the case of severe build-up, only long breaks, for example in long leys, can effectively reduce the populations.

Curiously enough, on many soils, cereal cyst eelworms decrease to small numbers even under continuous cropping with cereals. These soils evidently contain an enemy or competitor that seriously limits the eelworm's multiplication. The enemy or competitor is removed by formalin drenching (but not by some other soil fumigants), so that, after applying formalin, numbers may increase greatly to damaging levels, declining again as the effects of the formalin wear off.

AIR-BORNE LEAF DISEASES

Diseases affecting the leaves of cereals are important, as they can reduce yields by interfering with photosynthesis, the essential process which produces carbohydrates in the plant. If these diseases persist and develop within a maturing crop, they injure the flag-leaf and the ear itself; as these areas are sources of carbohydrates for grain-filling, this can result in poor yields.

The many foliar diseases include: yellow rust and black rust of wheat; mildew of oats, wheat and barley; leaf blotch of barley; and crown rust of oats. Foliar diseases that are air-borne can invade from external sources but the farmer can influence their spread, because they survive the winter on volunteer plants, whose destruction where possible can limit spread.

Yellow rust

Yellow rust (*Puccinia striiformis*) of wheat is an example of an air-borne leaf disease, and one that can reach epidemic proportions and infect early enough to decrease leaf efficiency greatly. The spores are air-borne, may travel many miles and enter the crop by a means over which there is no control. No reliable foliar sprays are available, though some farmers have tried lime-sulphur during epidemics with little effect.

Varieties of wheat show varying degrees of resistance. This can be short-lived, for the fungus can mutate, giving rise to new races or strains capable of infecting varieties hitherto immune. These races can be very specialised. Thus race 2B was first identified in 1954 when attacking Cappelle-Desprez, and race 55 was first reported in 1965 when so virulent on Opal spring wheat. The famous outbreak in 1966-67 on Rothwell Perdix involved race 60. These highly specialised wheat-infecting races do not affect barley or weed-grasses.

Outbreaks of yellow rust are usually troublesome after mild winters when the fungus can develop on volunteer plants in stubbles or winter corn. These can then act as centres of spread, particularly with warm, dry spells occurring in the summer. As the rust does not survive on seed or dead plants, it is important to control possible winter survival on living material. Volunteer plants from shed grain should be buried by ploughing, or sprayed with paraquat, before crops are drilled in adjoining fields. The use of varieties which show high resistance to known races is an important point.

Powdery mildew

Powdery mildew is a major source of yield loss in oats, wheat and barley. It too possesses a large number of virulent races, and has the capacity to overcome resistance by mutation; for example, this happened in Maris badger spring barley and Manod oats. Systematic fungicides applied as a seed dressing or foliage sprays give good control of powdery mildew in barley crop. Resistant varieties are helpful until they break down to another strain of the fungus.

Leaf blotch

Leaf blotch of barley can be likewise very important. Again, the value of stubble-cleaning and of avoiding winter varieties which may act as a bridge between the seasons is important.

WEED CONTROL IN ALL-CORN SYSTEMS

Good weed control in all-corn systems is of primary importance for many reasons. These include:

The reduction or elimination of competition which would reduce yields;
Easier harvesting, drying and storing of the grain (climbing weeds

such as bindweeds and cleavers, and weeds with heads such as poppies, sow thistles and mayweeds, are particularly troublesome in this respect);

Prevention of increase in the weeds by seeding, rhizomes, etc.;

Improvement in quality of the grain, especially for seed purposes, by elimination of weed seeds; these might be difficult or impossible to separate, e.g. cleavers, wild onion, wild oats, wild radish, etc.;

Reducing the risk of spread of pests and diseases; e.g. some weed grasses are hosts for cereal cyst eelworm; cleavers and wild oats are hosts for oat stem eelworm; and couch grass (*Agropyron* and *Agrostis gigantea*) and watergrass or creeping bent (*Agrostic stolonifera*) are alternative host for take-all and eyespot.

Annual broad-leaved weeds

Prior to the introduction of modern chemical weed control, cereals were regarded as the dirty crops in the rotation because they allowed weeds to grow and seed. Some chemicals such as sulphuric acid, copper sulphate, dinitro-phenols and cresols were used before the Second World War with very limited and variable results, but in 1941–42 the discovery and development of MCPA and 2,4-D was a great advance. These and the other salts of the chlorinated phenoxyacetic, phenoxybutyric and phenoxypropionic acids (translocated growth-regulating chemicals) effectively control most broad-leaved weeds in cereals.

When MCPA and 2,4-D were introduced, the commonest and most aggressive broad-leaved weeds in cereals were charlock (*Sinapis arvensis*), corn or field poppy (*Papaver rhœas*), runch or wild radish (*Raphanus raphanistrum*) and fat hen (*Chenopodium album*) and it was a very common sight to see cornfields covered with yellow and/or red flowers in summer.

The application of MCPA or 2,4-D as a spray or dust had a dramatic effect and easily killed these common weeds, but the chemicals had to be used for many years because of the large numbers of dormant weed seeds present in most fields. However, many people were convinced that the broad-leaved weed problem had been solved.

Then it was found that other weeds, neither killed nor severely checked by MCPA or 2,4-D, had an opportunity to grow and seed in a way which was not possible previously under a canopy of charlock. Cleavers (*Galium aparine*) and chickweed (*Stellaria media*) were the first to become troublesome, and the introduction of CMPP (mecoprop) and 2,36–TBA/MCPA mixture in 1956 to control these weeds as well as the "MCPA" weeds was very welcome.

When chickweed and cleavers were controlled by repeated use of mecoprop, another group, which so far had proved resistant, began to spread rapidly. These were the polygonums, the main ones being redshank (*Polygonum persicaria*), black bindweed (*P. convolvulus*) and knotgrass (*P. aviculare*). In 1961, another chemical, dichlorprop (2,4-DP), and also a mixture of dicamba and MCPA, were introduced to deal with this new problem.

As these weeds were controlled, the mayweed group became dominant. They are now being tackled by mixtures containing one of the following: ioxynil, bromoxynil and 2,3,6-TBA. Corn marigold (*Chrysanthemum segetum*) has proved a difficult weed to control on many farms.

The weed pattern in fields which have grown a lot of cereals during the past 20–30 years has changed. In fields which have only gone over to cereal production recently, the weeds may still be at the MCPA stage.

The phenol chemical DNOC has been available since 1942 (also dinoseb since 1950) for dealing with MCPA-resistant weeds. Their use is limited because of their poisonous nature and the risk to the sprayer-operators and to game birds, and they are not now much employed; dinoseb is sometimes used in crops undersown with clovers.

The butyric chemicals MCPB and 2,4-DB were developed for use in cereals undersown with clovers; the addition of benazolin widens the spectrum of weeds controlled, especially chickweed and cleavers.

Control of broad-leaved perennials

Perennial broad-leaved weeds are not so easily controlled as the annuals. The most troublesome are creeping or field thistle (*Cirsium arvense*), docks (*Rumex spp.*), field bindweed (*Convolvulus arvensis*) and sowthistle (*Sonchus arvensis*). The foliage of these plants usually emerges later than the normal spraying time.

Field bindweed is now becoming troublesome on many farms; it can be controlled by spraying the cereal two or three weeks before harvest with 2,4-D, the amount of wheel damage to the crop, especially if it is lodged, being negligible.

Monocotyledonous weeds

The most troublesome annual monocotyledonous weeds are:

Blackgrass (*Alopecurus myosuroides*), found mainly on clays and heavy loams where winter cereals are frequently grown; Wild oats: spring wild oats (*Avena fatua*), found throughout the country; Winter wild oats (*Avena ludoviciana*) which have a mainly south midlands and south-eastern distribution.

Wild oats are now found on most farms and are on the increase. If allowed to grow unhindered, they can seriously reduce yields. Both blackgrass and wild oats can be controlled by tri-allate applied in the seed-bed or barban applied in the seedling stage at a cost of about £3 per acre; this treatment may have to be repeated for many years because of the prolonged dormancy of many seeds in the soil and incomplete control by the herbicides.

Grass weeds

Rough-stalked meadow grass (*Poa trivialis*) is a troublesome weed in autumn-sown cereals in some areas. It may be controlled by seed-bed application of tri-allate.

The worst grass weeds associated with all-corn systems on most farms are:

Couch or twitch (*Agropyron repens* and *Agrostis gigantea*), spread by rhizomes (underground) and seed;

Onion couch (*Arrhenatherum elatius*), spread by bulbous bases and seed; and

Creeping bent or Watergrass (*Agrostis stolonifera*), spread by stolons (above ground) and seed.

These can become quickly established, especially where there is no broad-leaved weed competition. They can cause trouble by reducing yield, by making cultivations and harvesting more difficult, and by acting as hosts for take-all and eyespot fungoid diseases.

The main methods of control are by cultivations or by the use of chemicals.

Cultivations can be used as follows:

Either: drag the plants, including rhizomes, to the surface free of soil, so that they become desiccated and die. This is dependent on several weeks' dry weather for success.

Or: chop the plants into small pieces (e.g. with a rotary cultivator) which die by drying out, or else send up new shoots which can be buried by further cultivations. By burying new shoots every 2–3 weeks according to growing conditions, the weeds eventually become exhausted and die.

The cultivations method may be used as stubble-cleaning operations or in the spring prior to planting; but the latter holds up planting of the next crop and there is a risk of drying out the soil too much.

In extreme cases the field may be fallowed for a year but this can only be justified in very badly infested fields.

Various chemicals can be used to control these grass weeds; they are usually expensive and not always reliable.

TCA and EPTC (in some circumstances) should be worked into the soil to be taken up by the roots and shoots respectively—they are particularly effective on *Agropyron repens*.

Dalapon and aminotriazole, when used separately or mixed, are mainly absorbed through the leaves; they work best when the plants are growing actively and there is plenty of leaf to absorb the spray. Cultivations a few weeks before spraying stimulate leaf growth and improve the effects of the spray. Aminotriazole inhibits chlorophyll formation and the plants exhaust themselves by producing pink and white shoots. Dalapon causes deformed shoot growth which later dies back.

All the above herbicides cost about £6–£7 per acre at the recommended dosage. This is a compromise between what is needed to ensure a good kill and what is considered to be economic.

Paraquat desiccates the aerial parts of the plant. This is usually sufficiently damaging to kill *A. stolonifera*, but only checks the other perennial grasses. However, if the leaves are desiccated by spraying every few weeks, these grasses are weakened enough to be killed. Paraquat can be used in conjunction with cultivations to destroy regrowth as it occurs: e.g. cultivations in dry weather, and paraquat when wet conditions make cultivations difficult or impossible.

Combinations of the above chemicals are being tried to control rhizomatous grasses that occur in many types of soil and weather conditions.

Chemical weedkillers, used properly, can be very useful; but if used at the wrong time or at wrong rates or on wrong varieties, they may check the cereal, and this could result in disease organisms causing more damage. There is some evidence that damage by take-all may be increased in this way.

Break crops

Many farmers are now growing crops such as field beans and oil-seed rape as "break" crops, to reduce the risk of pest and disease build-up and to allow a wheat break to be introduced into a continuous barley system. This increases profits. The same cereal machinery can be used to grow and harvest these crops. Contracts are desirable when growing them.

Oats, because of their resistance to take-all and eyespot, are being taken as a break crop on some all-barley farms. They usually yield well, but are much more susceptible to root eelworm than barley or wheat, and so may increase cereal cyst eelworm.

Maize for silage and grain in south-eastern counties of Great Britain is being tried because of its freedom from the pests and diseases of wheat, barley and oats. In this respect maize is a good break crop, but harvesting is a serious problem in a normal or wet autumn, and special attachments for the combine are expensive.

Other break crops being grown on some farms are:

Dried peas—harvesting is now simplified by the use of diquat or sulphuric acid as desiccants before direct combining, and the crop is grown in districts other than the main area in the eastern counties.

Sugar-beet for seed—undersown in cereals, or sown after a fallow, and harvested the following September. To avoid risk of spreading virus yellows, this crop is best grown well away from the main sugar-beet areas.

Grasses for seed—Returns from this crop are variable, largely because good weather for harvesting is so essential.

Some extra machinery may be required for these crops.

Break crops may be introduced at a definite stage in a rotation, e.g. every fourth or seventh crop, or may be introduced when diseases or pests become serious on a particular field. In the latter case, a 2-year break would be highly desirable.

All the break crops mentioned allow couch to develop freely, so they should not be taken when a field is infested with couch because this weed carries on take-all and eyespot and nullifies the break effect. To be effective, break crops should be cleaning crops in every respect, or allow time for cleaning.

Problems of continuous cereals

Farmers who are successfully following a continuous cereal system are always alert for signs of a build-up of diseases, pests and weeds, and in this respect they are often assisted by soil tests carried out by the advisory services. A well-proven system is usually followed for keeping the land clean and fertile: for example, the Jeffes' system of repeated heavy discing and cultivating immediately after combining incorporates the chopped straw into the soil, and exhausts couch rhizomes by chopping and burying the developing green shoots or by desiccating on the surface.

Straw disposal can be a problem. If straw cannot be removed quickly or chopped and worked into the soil, then it is usually burnt. Burning helps to control eyespot, especially if all the stubble burns fiercely. If there is grass weed in the stubble, a small amount of paraquat will desiccate it in a few days so that a better burn is possible. If the straw and stubble are burnt every year, the organic matter content of loams and heavy soils becomes reduced, good soil structure is difficult to maintain, and so cultivations can become increasingly difficult and dependent on very good timing; this in turn may require increased labour and machinery. Chopped straw, worked into the soil at harvest time, quickly breaks down and may not require the extra nitrogen application recommended for straw ploughed in later.

The now popular floor-drying and storage systems for grain are much less demanding on labour at harvest time than continuous-drier systems, so labour can be released for stubble cleaning.

The recently introduced system of "slit-seeding" cereals into the previous stubble reduces labour and machinery costs, but does not allow for couch control and is not always successful in other respects. For some inexplicable reason it appears to result in a useful diminution of take-all.

Continuous cereal production, with its minor modifications, is an exhausting system of farming. Nutrient losses must be made good in the fertiliser applied. 40–60 units per acre* of phosphate and potash will be required each year, and 80–100 units of nitrogen (or perhaps more on soils which leach readily). Combine drilling of fertiliser for cereals is still a common practice, but is being replaced by broadcasting on some farms with satisfactory results in most cases. Mechanisation to reduce labour requirements and bulk handling of fertiliser are the main reasons for the change in methods of application.

* 1 unit = one hundredth of one hundredweight = 1·12 lb

INTENSIVE ALL-ARABLE SYSTEMS

Those all-arable systems which include a large proportion of high-return crops such as potatoes, sugar-beet, carrots, vining peas, celery, brussels sprouts, etc. are found mainly on the rich deep soils of the eastern parts of Great Britain. These include especially the Fen area around the Wash, and also such areas as the warpland soils of the lower Trent valley, parts of Essex, the Lothian area of Scotland, S. W. Lancashire, the Isle of Thanet and some coastal areas in Devon, Cornwall and Pembrokeshire.

These intensive cropping systems make heavy demands on labour and machinery and in many cases are dependent on casual labour. However, in recent years there have been remarkable developments in mechanisation, giving better seed-bed preparation, automatic planting of most crops, precision seeders and monogerm seed (which more or less eliminate sugar-beet singling), herbicides to control weeds and fully mechanised harvesting. Also, the increased size of tractors and implements has further reduced labour requirements.

The main limiting factor controlling the production of these high-value crops in close succession is the risk of pests and diseases, especially cyst eelworms and, on light sand, root ectoparasitic eelworms. To avoid a build-up of these pests and, sometimes, to comply with quota and contract requirements, a break of at least 3 years between successive crops is required. Monoculture of cash "root" and vegetable crops is therefore very unusual. An exception is the early potato crop as grown in parts of Ayrshire, Pembrokeshire, Devon and Cornwall. This can be grown for long periods in successive years provided the crop is lifted early, which prevents most cyst eelworms present from completing their life cycle and increasing the population excessively.

Intensive all-arable systems require much capital expenditure on machinery and sometimes on buildings, e.g. for potato chitting and storage of ware. Seed and fertiliser costs are large. Labour requirements are high. Chemicals for weed, pest and disease control are also expensive. The crops carry much risk but the returns are usually high and profits can sometimes be considerable, especially in favourable seasons and provided that management is good—especially with regard to efficient use of labour and machines.

If looked at from the gross margin analysis point of view (see Chapter

12), it is usually found that direct costs, i.e. items such as seed, fertiliser, sprays, casual labour, etc. and sales are satisfactory because good husbandry methods are followed. However, in spite of this, the profits and returns on capital may be small. There may be many reasons for this, but in most cases the reason is that labour and/or machinery costs are too high because they are used inefficiently. Altering the acreages of the various crops to use permanent staff fully throughout the year, after making allowances for any reliable casual labour which may be available and for suitable mechanisation at peak periods, usually reduces the fixed costs or increases sales sufficiently to give adequate return on capital invested and a good profit.

Crops and rotations

A high standard of skill in field operations is required for intensive all-arable systems. On the heavier soils a considerable amount of patience is also required during seed-bed preparation.

Many of the crops produced on intensive arable farms are now grown on contract and have to be graded to a high standard. Some are subject to control throughout the season from planting to harvest, e.g. vining peas and green beans; they must be planted at carefully determined intervals, so that the factories will receive a steady supply during the harvesting period. Little sugar-beet is now grown in the south and south-west of Great Britain because of the long distance to the nearest factory.

Sugar-beet is deep-rooted, and is usually preferred to potatoes on soils which are likely to dry out quickly and where irrigation is impossible. If irrigation is possible all these intensive crops will probably benefit in most years, in yield and quality, in the eastern part of the country.

Apart from busy periods, such as planting and harvesting and any period after unfavourable weather, weekend working can be avoided: the 7-day week associated with livestock enterprises is unnecessary.

Where intensive cropping is practised the soils are usually rich in nutrients because much fertiliser is applied, but organic matter has to be carefully watched. When no farmyard manure is available, it is usually desirable to chop and return the straw from the cereal crops which are grown.

On many arable farms the cropping is predominantly cereals, but cash root crops such as sugar-beet or potatoes are grown to a limited extent to increase profits. The acreage limit may be determined by such factors as the number of fields suitable for these crops, or shortage of labour or capital. When sugar-beet follows barley on coarse sandy soils in eastern England and elsewhere, it may be stunted by stubby root (Trichodorus spp.) and needle eelworms (Longidorus spp.). These destroy lateral rootlets, prevent nutrient uptake and cause a condition known as Docking Disorder. Recent experiments suggest that the condition can be controlled by drilling the sugar-beet seed into rows pre-treated with small amounts of DD or Telone nematicide, and the method is now being applied commercially. These eelworms injure several other crops, e.g. carrots and sometimes even spring-sown grasses.

Rigid crop rotations, where a number of different crops are grown in a fixed sequence, are not so common nowadays. The same acreage of each crop may be grown every year to simplify labour and machinery requirements and to conform with contract or quota limitations, but the sequence of cropping may be varied in different fields and only the principles of sound rotations considered when planning the overall cropping each year.

Sometimes grass will compete with arable crops; some land is only suitable for grass. Grass as a crop on a wide range of soil types is dealt with in the next chapter.

Chapter 4

ALL-GRASS SYSTEMS
INTRODUCTION

Grass is the most important British crop, and on many farms is the only one grown. These all-grass farms, or farms where utilisation of grassland is the major enterprise, are widely distributed, though many may consider them to be common only in the traditional fattening areas of the Midlands, the dairying areas of Cheshire and Somerset, the rearing farms of the English-Welsh border and the marginal and hill sheep farming areas of Scotland, Wales and north and south-west England. Very often they are found in those areas where a combination of high rainfall and soil type are not conducive to arable cropping and where the abundance of grass growth is able to support livestock production at profitable levels.

Whilst the severity of climate will be a major factor in the development of grassland farms in hill country, it must be realised that in lowland areas the policy may be determined by farm size. For example, where 100 acres or less is the only land resource, then intensive stocking with dairy cows is often the only way of achieving high profitability per acre. Grass farms support a variety of livestock enterprises.

USE OF PERMANENT GRASSLAND

On all-grass and livestock farms the use of permanent pasture and long leys rather than short temporary leys is popular, for the former have several advantages. Infrequent re-seeding of pastures reduces production costs and the loss of acreage during re-seeding is reduced, Frequent re-seeding, especially in spring, gives a loss of valuable high-yielding flush grass, though late summer re-seeding reduces this effect. Although spring re-seeding can give very high quality herbage as maiden seeds to be grazed in June and July, and so extend the period of quality grass, it is more expensive and of lower overall production than is found in a run of years with high-quality permanent pasture. The latter is also more resistant to poaching and intensive grazing pressure, and in moister areas and on heavy land minimises damage from high stocking density.

A further advantage of long-term grassland is that its somewhat shorter season of growth suits many heavy land areas where early spring grazing may be limited by the effects of excessive poaching. Indeed in many grass areas, soil drainage is probably the factor needing most attention for higher grassland output.

Permanent pastures of high quality

Permanent pasture is of varying quality and productivity, and much of this is due to management as well as other environmental factors, especially soil fertility. The best swards are those dominated by indigenous perennial ryegrass (*Lolium perenne*), and although of limited total acreage they are widespread in the British Isles. First-class pastures contain over 30 per cent ground cover of perennial ryegrass, and are often referred to as "fatting" pastures.

Perhaps the most famous ones in England are found in Leicestershire and Northamptonshire where they overlie lias and boulder clay, e.g. along the Welland valley, and reach their best development in the neighbourhood of Market Harborough. These pastures are usually stocked with heavy beef stores under the system known as "set-stocking", which means that apart from adjustments to stocking rates for seasonal growth of grass, the animals are left to graze in the same fields throughout the grazing season. A second famous area is found in the marine silts of the Romney Marsh in Kent where intensive set-stocking by sheep and lambs often reaches 12 ewes per acre.

Other centres are well known, such as the feeding grounds of the Northumberland coastal plain distributed from Alnwick to near Berwick-on-Tweed. Here the soils are medium to heavy clays and the best pastures are probably in the Belford district. In Lincolnshire the pastures on coastal marsh alluvium, particularly near Wainfleet, are also an example of high-class permanent pasture. Another centre occurs in Dorset, where on the alluvium overlying gravel in the Stour Valley the pastures are particularly good near

Manston and Hammoon. Other English examples occur locally on a restricted basis along various river valleys as in Herefordshire, Devon and the Thames valley.

These swards not only contain perennial ryegrass but other useful grasses with white clover (*Trifolium repens*) the most important legume. Some twenty-five species are commonly found, though typical dicotyledonous weeds such as dandelions and buttercup are normally poorly developed. The influence of perennial ryegrass on feeding value has long been appreciated, and Dr. William Davies classified swards as 1st grade with 30% and over of this species, 2nd grade 15–30%, 3rd grade 12–15% and 4th grade when ryegrass was absent or nominal.

Welsh examples are limited, though some of high regard occur on the seaward side of the estuary flats of Monmouthshire between Cardiff and the River Wye and again on the alluvial riverside pastures of the Dee in Denbighshire and Flintshire. In Scotland with its long tradition of alternate husbandry there is little "fattening" pasture of the type found in England. However, on some heavy soils, old leys have developed into such pastures and good examples are found in the Cruden Bay area, north of Aberdeen, where first-rate beef is fattened. In Northern Ireland the ley farming tradition is probably responsible once more for the absence of first-class traditional old pasture but the lack of inherently fertile soils may also be important. Some areas are found in Eire but only those in the Dunshoughlin district of County Meath have an outstanding reputation. The best Irish permanent pastures are similar in type, management and production to those of the English Midlands.

Permanent pastures of lower quality

Much of the permanent pasture found on grassland farms is far from the quality of these high-class permanent ryegrass swards, and these poorer communities are dominated by species such as bent grasses (*Agrostis tenuis* and *A. stolonifera*) and red fescue (*Festuca rubra*), together with a wider range of associated weed species. After proper treatment to correct deficiencies of lime, phosphate and potash, the application of nitrogen fertilisers stimulates growth, and provided a system of rotational grazing is practised, increasing livestock output results.

The consequent effects of defoliation and gradual fertility build-up result in gradual and then quick changes in botanical composition. Residual perennial ryegrass spreads and assumes dominance, and species such as rough stalked meadow grass (*Poa trivialis*) and smooth stalked meadow grass (*P. pratensis*) increase. In this way poorer pastures can be improved, and a variety of surface cultivations can be adopted to encourage and speed the changes. Other pastures may contain rushes, and here the re-opening of drains, surface cultivations and herbicides, where appropriate, will all aid improvement.

Fattening

The high quality fattening pastures are due in the main to fertile soil and good grassland husbandry, whilst the presence of a high water-table provides a good growth of grass into mid-season. This has permitted a traditional beef fattening system that has, in some areas, remained unaltered for many generations, though this is now being challenged by the profitability of wheat growing, so that the plough is making inroads into these all-grass farm areas. The management is to set-stock according to available herbage at a beast or more to the acre. Stock are put on in spring and the rate of stocking adjusted according to necessity with an overall aim of fattening the cattle by the end of the summer without cake feeding and with as little cost as possible.

Fertilisers are usually limited to periodic dressings of basic slag or potassic supers; whilst nitrogenous fertilisers, although they can increase stocking rate, are not normally used.

Liveweight gains of over 200 lb per acre are usual for this short spring and summer fattening season. Much of the success of fattening pasture enterprises rests on the farmer's ability to buy and match his stock well and to sell at the right time. One interesting point about these pastures is that the natural selection and the grazing management have produced a perennial ryegrass of a persistent, leafy and dense-tillering type that has been used in the breeding of varieties such as Aberystwyth S.23, and in the production of seed crops of local indigenous forms such as Kent Perennial Ryegrass.

DAIRYING

Stocking rate and fertilisers

Dairying, where practicable, can be the most profitable livestock system of grassland utilisation. On many farms this has been achieved through intensifying the annual stocking rate with a general objective of maintaining a cow to the acre for all its grassland requirements, though this is not always realised. In this way smaller farms have been able by the use of heavy fertiliser dressings and rotational grazing allied to good stockmanship, to maintain profitability in the face of rising costs. The use of nitrogenous fertilisers has been a key factor with annual applications reaching up to 300 units per acre and giving economic returns on grazed areas. With several cuts of grass from silage blocks good responses can be obtained from up to 400 units per acre. On some farms the grazed areas have exceeded the 400-unit level, but this is more usually about 250 units per acre. In all situations there is need for attention to detailed requirements of phosphate, potash and lime, though potash should not be applied on the grazing areas in spring due to the risk of hypomagnesaemia. The use of 40–60 units of nitrogen before the first and after each subsequent grazing has brought to grass production a precision which has given greater confidence in supporting high stocking rates, and demands for yields in excess of 10,000 lb of dry matter/acre/season have in this way been met.

The amount of phosphatic and potassic fertilisers applied varies with the soil's original levels of these elements and the management used. On grazed areas, with a high return of plant food in the dung and urine, 25 units of phosphate and 50 units of potash per acre may be adequate, whereas on cut swards the requirements per ton of dry matter cut are in the region of 20 units of phosphate and 30 units of potash. With swards giving as much as 4 tons of dry matter per acre per season, there is clearly need for high rates of application. Since beef and sheep enterprises do not export from the farm such high quantities of nutrients as the dairy cow, smaller amounts of phosphate and potash are needed, and 25 units of each/acre/annum are usually adequate in their case.

Method of grazing

As herbage yields have risen on dairy farms, there has been a greater reliance on rotational grazing either by paddocks or strip-grazing employing a back fence. It is worth noting that intensive sheep enterprises have used forward creeps in the same way, and semi-intensive beef and even suckler herds have been grazed increasingly in paddock systems. On dairy farms the overall grazing precision achieved has led to a possible 3-week grazing cycle, often based on a minimum of 21 1-day paddocks, with the paddock size being determined by an average allocation of 0·02/acre per cow per day's grazing.

Some farmers prefer 2–3 day paddocks, sometimes using two lanes by splitting their herd into high yielders that receive the first bite of herbage then followed by the low yielders. Another system developed at Wye College (University of London) employs four large paddocks, each strip-grazed for seven days and giving a 28-day cycle. Whilst strip-grazing has been somewhat less publicised and has been a little out of favour because of the necessity for repeated movement of fences every day, it is an efficient method and very popular with farmers requiring a high degree of utilisation. It also affords greater flexibility and is more easily adjusted to the varying rates of grass growth during the season, especially the high rates found in the spring flush: if any surplus arises at that time it may be made into silage or hay rather than allow the grass to grow out of hand.

Main approach

Thus, in effect, the farmer has, through the use of nitrogenous fertilisers, been intensifying production within the natural rhythm of herbage plants and he has tended to avoid using exceptionally sophisticated mixtures and practices for out-of-season keep. It is common to see 150 units and over per acre of nitrogen applied by the end of May on grazed areas in order to obtain the high response of the spring flush that is frequently twice that found at other times of the year, and in the order of 30–40 lb of dry matter per lb of nitrogen applied. A late or slow season will modify rates of application, but the overall effect has not only been to produce more grass but to increase growth in the post-flowering, slow-growing period, possibly due to better root development, and to give a narrowing of the traditional summer gap period.

The high rates of nitrogen also increase the vigour and competitiveness of the grasses which in turn reduce the cover of white clover. In some cases this does not happen, and in others the clover increases in later years due possibly to the greater availability of nitrogen as fertility increases. Nitrogenous fertilisers promote quick grass recovery, and sometimes non-protein nitrogen in the form of nitrates may accumulate temporarily in the herbage; when grazed too soon this can have a depressive effect on the milking cow. It is important not to make too quick a return to top-dressed pastures, and where a 2-week cycle is being approached great care should be exercised.

Dung spoilage and poaching

It might be thought that with grazing occurring 6–9 times per season on many paddocks, there would be a massive accumulation of dung, consequent spoilage and a decrease of grazing area as the season proceeds. However, this has not resulted, for it appears that as the grass–animal–soil cycle has intensified, so has the rate of dung decomposition. Mineralisation by bacteria re-cycles nitrogen into the soil with an increasing availability that, on some farms, has led to the need for less nitrogenous fertiliser application. A feature of very intensive grazing régimes using high nitrogen rates is that weeds tend to be reduced to minor proportions due to the intense plant competition and defoliation. In some areas however, broad-leaved dock has become prevalent and the timely development of herbicides such as Asulox could be very useful for its control.

With stocking density increasing to levels of less than one acre of grassland per cow, the possibilities of excessive spoilage and poaching and sward destruction become more real, and some farmers have turned to zero grazing as a means of maintaining intensification. This uses the same technique of grass production but cuts and carts the grass to yarded cows. It requires greater organisation and application, and has not been popular with farmers, who have rather tended to meet the challenge by superb grazing management and splitting their herds to give two smaller units using separate grazing areas. In this way sheer physical grassland damage, as at the entry to paddocks, is reduced.

Use of leys

Whilst considering the effect of nitrogenous fertiliser on grassland production it is worthwhile taking into account the type of sward involved. Surveys in Cheshire have shown that the response of permanent pastures to nitrogen is more or less linear, and they were found to give almost as much utilised starch equivalent output as leys at application of from 60 to over 160 units of nitrogen/acre/annum. Generally the response of permanent pasture to nitrogen is the same as on leys, though the total yields might be slightly lower. The practical point is that well-fertilised and well-managed permanent pastures are capable of very high production, for some of the "natural" grasses such as *Agrostis* are very responsive to nitrogen. However, on poor matted swards response can be slow due to bacteria decomposing the mat and using much of the nitrogen. Then, aerating the mat by surface cultivations to improve the rate of decomposition is important.

It is necessary from time to time to plough and re-seed pastures, and direct re-seeding required to give the rapid return to productive grassland needed on all-grass farms. In longer leys the use of persistent leafy pasture varieties of perennial ryegrass is highly desirable. These varieties are persistent under rotational grazing and heavy trampling, and they are responsive to nitrogen and high fertility. They also offer levels of organic matter digestibility as high as, if not higher than, other species or varieties, and ensure high herbage intake by livestock. Mixtures including timothy are popular; and whilst those based on timothy, meadow fescue and white clover are often quite outstanding, particularly on heavy soils, they do not have the high recovery rate (found in ryegrasses) that is needed on intensive dairy systems. Leys containing a high quantity of cocksfoot are not desirable as this species is generally lower in digestibility than other species.

Italian ryegrass has no great advantages over perennial ryegrass, and its lack of persistency and need for frequent re-seeding is unpopular on the all-grass farm due to the higher costs, though it plays a much more important role on mixed farms where shorter leys are often required. Although white clover can have a low survival in high-nitrogen systems due to the competition of grasses for nutrients and moisture, its feed value is great on account of its high digestibility, mineral richness, protein and vitamin content. Whilst some advocate its exclusion from so-called "high-nitrogen" mixtures, its feed value makes even a small contribution worthwhile, especially as animals seem to have higher herbage intake and fatten more effectively on clovery swards.

On intensive grass farms with irrigation, white clover has a greater survival; and the same applies, in general, to farms in high-rainfall areas. White clover also survives better under intensive rotational grazing systems than it does under intensive silage production. At less intensive levels of management, white clover's nitrogen-fixing capacity increases production and may, on swards with a cover of 35 per cent clover and over, yield 100–200 lb of nitrogen/acre/annum. However, white clover starts growth at a soil temperature of about 9°C (48°F) and is later than many grasses that can start at 5·5°C (42°F) soil temperature, and so it is unable to transfer nitrogen to grasses in early spring. It should be noted however that its nitrogen transfer lacks the precision of growth stimulated by an application of artificial nitrogen.

Red clover is a traditional hay plant that is normally used in short leys. However, in the longer leys used for conservation, the persistent late-flowering varieties can be sown to improve the protein content of silage cut in the first few seasons. Lucerne is not prominent on all-grass grazing farms, but occasionally, where high quality hay is produced by artificial drying and used for winter milk production, it is sown with a small quantity of a companion grass such as timothy in 3- to 4-year leys.

The decision to plough and re-seed is often controversial, but it is fair to say that ley mixtures offer a means of extending the somewhat shorter season of permanent pasture. Even if this is not required, perennial ryegrass leys with their high digestibility values, which in bred varieties can exceed those found in indigenous types, also offer exceptional response to nitrogenous fertilisers and allow livestock to grow and produce effectively. They also provide massive single cuts, and where long leys are used, the costs involved in re-seeding are minimised.

Herbage conservation is important on grassland farms and a fuller account appears in Chapter 3. Silage production from special cutting blocks and leys has been well-developed on intensive farms, whilst some practise silage making and grazing alternately. The silage aftermaths can give very useful leafy herbage in late June and early July and are very suited to maintaining milk output. The silage itself plays an important role in winter milk production, and can also play a vital one on semi-intensive beef-fattening systems to maintain yarded suckling cows and to help finish beef animals after summer grass.

Haymaking is of traditional popularity everywhere. On hill farms, hay is a more useful product than silage, as it is more easily transported about the open grazing during critical snow-bound periods.

Seeds mixtures

Seeds mixtures are very variable, but those for high production must include plants that are of high digestibility and therefore do not restrict the animal's herbage intake. This demand, together with those of simplified management and the need for seasonal predictability of growth pattern, heading and feeding value, have led to the advocacy of progressively simpler mixtures capable of intensive production as long leys for all classes of livestock are tabulated below:

		lb/acre		lb/acre
Perennial ryegrass	S.23	18	Sceempter pasture ryegrass	18
White clover	S.100	1½	White clover Kersey or Blanca (RNP)	1½
,, ,,	S.184	½	White clover S.184	½
		——		——
		20		20
		——		——

Suitable long-term mixtures with earlier production could include S.24 perennial ryegrass (or S.321 in milder areas) and possibly timothy. Here are two examples:

		lb/acre	lb/acre
Perennial ryegrass	S.24	8	8
,, ,,	S.23	10	6
Timothy	S.48	—	4
White clover	S.100	1½	1½
,, ,,	S.184	½	½
		20	20

A somewhat later spring production could be obtained from the following mixture, which could be used as a 4–7 year ley:

Timothy	S.51	3
,,	S.48	3
Meadow fescue	S.215	8
,, ,,	Mommersteeg pasture	4
White clover	S.100	1½
,, ,,	S.184	½
		20

Leys for silage could be based on early perennial ryegrasses, which tend to hold their digestibility for a while after heading, and on red clover to improve protein levels in the early years. The following is a suitable mixture:

Perennial ryegrass	S.24	20
Red clover	S.123 or Hungaropoly (tetraploid)	4
White clover	Kersey	2
		26

Seed rates vary considerably from area to area; those quoted are lower than used in the north of England and Scotland.

FEEDING AND REARING FARMS

A third type of grassland is found in many of the grassland areas of the English-Welsh border and in the areas approaching a marginal land classification that is best described as Rearing or Feeding. A suckler herd, sheep flock and fattening enterprise can be found on these farms. The needs of the beef and sheep grazing systems and of conservation can fit in together, for intensive sheep stocking does best on swards free of sheep the previous season, and so it is possible to alternate with cattle grazing or conservation. Both the sheep flock and suckler herd are usually set-stocked, and mixed stocking occurs on some farms and demands considerable skill in obtaining the right balance.

Intensification of the stock enterprise is not easy in some of the harsher conditions but paddock-grazing of cows and calves and forward creep grazing of sheep can be undertaken. In addition, autumn-born calves, after their winter rearing, can be taken through on a semi-intensive paddock system, with 2 or more calves per acre of grass, employing the same basic principles of intensive management already discussed. Further fattening indoors can produce 18-month beef, though suitable buildings are not always available for this method of fattening. The grassland on these farms can be very mixed in type, including leys and permanent pasture and varying in quality from perennial ryegrass to *Agrostis* dominant swards. The latter are frequently the result of necessary seasonal mis-management resulting from excessive winter stocking or uneven summer grazing when acting as reserve areas to the main pastures, but are of course improvable.

Hill farms

Hill farms have a wide variety of grassland types. Because they are affected by extreme features of climate, high precipitation and leaching, they have acid soils; their season is shortened by the effects of elevation on temperature. These features limit the choice of stock and enterprises. There is a wide variation in the grasslands of enclosed areas, for these are always subject to the pressures of winter stocking and summer haymaking—both of which on a long-term basis result in *Agrostis* dominance. Dairy cow grazing and winter keep for sheep can be obtained from enclosed areas and some stores may be kept.

Maintaining fertility is chiefly dependent on the use of lime, phosphates and potash, but on open hill grazings fertilising is seldom practised, although by its use the production and maintenance of perennial ryegrass pastures are possible—but usually at a prohibitive cost. On well-drained slopes, particularly if south-facing, bent-fescue pastures are frequent and much favoured by sheep. Poorly-drained level areas are often colonised by flying bent (*Molinia*), whereas mat grass (*Nardus*) may occur if drainage is reasonable. In addition heather-dominant areas may be found, and provide valuable winter feed where rotational block burning is practiced to maintain the heather in a short "green" condition. All these swards are of lower potential value than bent-fescue areas. A major problem is to obtain a balanced seasonal and complementary use of each type, and this can call for exceptional shepherding skill.

The problem of bracken (*Pteridium aquilinum*) colonising good, well-drained land is very real in many areas, especially as its poisonous properties make it a dangerous fodder and its shading capacity eliminates desirable species.

It is important that hill grazings are not overstocked, as excessive grazing leads to sward deterioration and to poor nutrition of the ewe, with consequent poor lambings and reduced overall productivity. In particular, attention must be given to the nutrition of the ewe in winter, especially on the open hill, if higher productivity is to be achieved. The use of stores or suckler herds on hill grazings is valuable, as the cattle are able to graze the more fibrous grasses such as *Nardus* and help maintain desirable species.

The possibility of improving hill grazings in small and suitable blocks is essentially an economic matter, and dependent on the basic principles already discussed for permanent pastures. Additionally there is the need for attention to cultivations that aerate the surface mat and promote decomposition of accumulated organic matter. Observations on sites used by sheep as shelter areas show how the gradual build-up of soil fertility from animal faeces can alter the botanical composition; these studies indicate the great potentialities that hill grazings represent. Much experimental work carried out by Stapledon in Wales and elsewhere in the 1920's and 1930's, and at the present experimental husbandry farms and by individual farmers, has shown that these potentialities can be realised.

SUMMARY

From this general and brief consideration of farms which are predominantly grassland enterprises, it is clear that there has been a considerable advance in technique and output during recent times. Whether it is in the use of permanent pastures or long leys, the principles are basically the same: namely the use of rotational grazing systems that allow a period of recovery to the grazed sward, supported by the use of balanced fertiliser applications to maintain or improve soil fertility. On dairy farms there has been most intensification. Beef production offers less scope for more than semi-intensification, and hill-land livestock enterprises require more attention to the details of animal husbandry and capital to achieve higher outputs.

Chapter 5

MIXED ARABLE AND GRASS SYSTEMS

Throughout the British Isles, it is usual to find grass and arable crops on the same soil types. Mixed arable and grass systems of farming can be conveniently discussed as:

(1) Dairying and arable farms;
(2) Dairying, fattening and arable farms;
(3) Fattening and arable farms.

Although these systems will be seen in almost every county of the British Isles, they can be broadly regionalised as follows.

(1) Dairying and arable farms are found in most of the Midlands and south central England, north-west England, central and south Wales, central eastern Scotland and Northern Ireland.

(2) Dairying and fattening farms with arable are found in parts of south-west England, south central England, the north Midlands, parts of Wales—particularly Anglesey and South Wales—parts of eastern Scotland and in Northern Ireland.

(3) Fattening and arable farms occur in the east Midlands, north-east England, parts of central, southern and eastern Wales and eastern Scotland.

DAIRYING AND ARABLE FARMS

The alternate husbandry system

With some exceptions, farming is based on the alternate husbandry system. This can be described as follows: a field is put down to grass as a ley to be grazed or conserved as fodder for normally a pre-determined number of years, and the fertility built up by the ley is eventually cashed in with a series of arable crops.

It is true that the ley will increase the organic matter content of the soil, but experiments in Great Britain over the last 30 years starting at Woburn (a sandy loam soil) and then at Rothamsted (a clay loam) and now at some of the Experimental Husbandry Farms of the Ministry of Agriculture, Fisheries and Food (on different soil types) suggest that the alternate husbandry system may not be as beneficial as at one time thought. Many would query the need for these experiments because it has for long been common practice to "take the plough around the farm", and the benefits are still considered too obvious for the policy to be questioned. Furthermore, until recently grants given by the Ministry to encourage farmers to plough up grass of more than three years' duration have in practice endorsed even more the principle of the alternate husbandry system.

The original aim of the grant was to encourage the ploughing-up of old pastures—pastures which very often had not been ploughed for 30 years or more. There is no doubt that the following arable crops made the field far more productive, and when eventually it was put back to grass or was for some reason direct-reseeded, the resulting ley was certainly more productive than the original old pasture. In many cases, better productivity has been attributed solely to the ley, when in fact far better treatment in terms of fertilisers and management has been given to it than was the practice with the old pasture.

Putting an arable field down to grass will not increase organic matter to any great extent. For example, at Rothamsted, an old arable soil seeded down to grass increased the organic matter content by 1·4 per cent in 12 years, and at High Mowthorpe Experimental Husbandry Farm (medium loam over chalk) where a ley was maintained for 6 years, soil organic matter was increased from 3·8 to 4·2 per cent—only approximately one-tenth.

In the various ley/arable rotation experiments, crop yields following 3-year and longer leys have been shown to be very little higher than the same crops following continuous arable. Even on light, sandy loam soils at Woburn in Bedfordshire where a ley might have been expected to improve soil structure, the same picture emerged. In many cases the slightly lower yield following arable cropping could quite easily be remedied by a little extra nitrogen.

One of the advantages of the alternate husbandry system is that the ley can help to reduce some, but not all, soil-borne pests and diseases which may be built up by all-arable cropping. Precautions can now be taken to control many of these pests and diseases other than by putting the field down to grass.

This is not to say that the ley or for that matter the alternate husbandry system will necessarily now be abandoned, adopting instead permanent grass on one part of the farm, and continuous arable on the remainder. There is much to be said in favour of the latter approach, with the permanent grass and dairy nearest the buildings, and all the arable cropping further away from them. Simplification of management could be one great advantage, and provided the permanent pasture is good permanent pasture, and is afforded the same management as would be given the ley, there is no reason why production from the grass should not be as good as, if not better than the ley (see Chapter 4). On some mixed dairying and arable farms this trend is now developing. Longer leys are being used within easy reach of the dairy herd, and shorter leys form the break crop, albeit in many cases a poor one, in a run of cereals. Of course, long leys and permanent pastures are to be found away from the buildings. With a milking bail during the summer these grass fields can be properly utilised by the herd.

Conservation

In addition to grazing, grass makes an important contribution to the winter feed for the dairy herd. Silage is not yet universally popular with the British dairy farmer. About 80 per cent of the total dry matter of all the herbage conserved for feeding in the United Kingdom is made into hay. Various reasons can be given for this fact, but none of them are entirely valid. Compared with hay, silage is unpleasant to handle, but where a self-feed system is adopted, although it is not always possible, this problem is overcome. There can be excessive wastage with silage amounting to 40 per cent loss of digestible nutrients, but with hay it can be considerably higher, although it is often less apparent.

Silage-making

It is traditional to make hay on dairy farms, and it would be relevant to say that silage has, to an extent, been the victim of its own propaganda campaigns which have appeared every few years in the last 25 years. Each campaign has brought out different advice as to how to make the best type of silage and how to achieve the correct fermentation. The latter has been the subject of much discussion. This has certainly led to confusion in the mind of the farmer, who has, in consequence, tended to keep to something he knows about—hay. In addition, changing from hay to silage can involve increased capital investment in machinery and buildings.

The discussion about the correct kind of fermentation now appears to have resolved itself into the so-called "hot" and "cold" fermentation methods. This refers to temperature, and there is now no doubt that a good lactic silage can be quite easily produced whether the temperature is at blood heat (about 37°C) or below 16°C. Many farmers are making cold silage; logically this must be right, because at the lower temperature there will be less breakdown of the carbohydrates and therefore less loss of dry matter, and the digestibility of the material will be higher. There are other farmers who like a higher temperature because they feel that there will be less risk of butyric bacteria (producing butyric acid silage) dominating the environment. However, if precautions are taken by ensiling the green crop in as dry a condition as possible, and/or using an additive to assist the lactic bacteria, there should be little risk of the wrong sort of silage being made.

Many farmers now wilt the cut crop in the field prior to ensiling. As mentioned, a drier crop means better conditions for the lactic bacteria when making cold silage, and in addition it is an absolutely essential prerequisite for high dry matter silage especially when made in a tower silo. Wilting will also cut down on effluent loss.

If cold silage is to be made with wilted material, difficulty can arise in securing the necessary compaction in the silo to prevent over-heating. This can be overcome by laceration of the wilted grass, a process which makes compaction much easier and which is now widely adopted. A direct-cut leafy grass crop can produce an effluent discharge of as much as 100 gallons per ton of grass carted from the field. This can be reduced to virtually nothing if the cut crop is wilted to 25 per cent dry matter (DM). With self-feed silage, the cow eats up to 100 lb per day, and obviously the higher the dry matter the higher the nutrient value.

There are disadvantages in wilting, not least of which is the fact that dry matter loss in the field can be quite high; but on balance it is preferable, although it does introduce a new factor into silage-making, i.e. dependence on the weather.

It has been generally accepted that high dry matter silage is less wasteful compared with the sealed clamp system, and furthermore it can normally be more easily handled when finally fed, although it is a less flexible system. There are farmers, who have been making high dry matter silage in towers for a number of years. A closer examination of the two systems can reveal whether the extra efficiency attributed to the tower system justifies the capital investment involved.

There are various ways of making silage in a clamp, but many farmers have sufficient evidence which shows that the sealed clamp (Dorset Wedge) method is the most satisfactory. This works on the same principle as silage made in the tower. The aim is to kill the plant cells in the shortest possible time. This means that air must be prevented from entering the mass of green crop in the clamp. The oxygen originally present in the cut crop is soon used up, respiration ceases and the plant cells die. Thus respiration losses are kept to a minimum. The silo is filled in the form of a wedge, starting at one end, and extending until eventually the whole floor is covered. At the end of each day's filling the crop is covered with a plastic sheet to trap the air in the silo and to prevent any more air from entering.

It must be assumed that silage from the tower will, when fed, be capable of not only maintaining the animal, but also of making a sizeable contribution to the production ration. This will necessitate wilting to between 35 and 50 per cent dry matter in the field prior to ensiling, and this is the weak link in the system, for the following two reasons.

(1) The weather may not be suitable for wilting to the moisture content required, and ensiling will have to be postponed until the crop can be properly wilted. Postponement, from what can be considered as the most practical time in many parts of the country to ensile grass, i.e. the last 2 weeks in May, will mean that the digestibility of the majority of the grass species will quite quickly decline. This results in a crop of lower feeding value being eventually ensiled and subsequently fed.

(2) Field losses expressed as a percentage loss of dry matter can be quite high. Obviously the longer the period in the field, and the more handling the crop receives, the greater the loss (see Table 5.1.).

Effluent loss, as has been mentioned, can be reduced to virtually nothing with wilting, and losses through spoilage, such as a black and slimy layer which can occur at the top and at the sides of the silo, are largely management problems.

Table 5.2 compares the losses from high dry matter tower silage and sealed clamp silage.

Thus overall conservation efficiency for the two systems differs by only about 3 per cent. So often it has been claimed that total losses with the tower are 20 per cent less than with the clamp. What has not been

Table 5.1
Average field loss of dry matter due to wilting

Dry matter of silage (%)	Loss of dry matter in field (%)
25–30	10
30–40	12–13
Over 40	15–16

Table 5.2
Average loss in silage-making (dry matter %)

	Loss in field	Loss in store			Total
		fermentation	effluent	spoilage	
High DM tower silage (45% DM)	15	8	0	0	23
Sealed clamp silage (25% DM)	10	12	1	3	26

fully appreciated is the field loss when the crop is being wilted, and with young leafy grass in showery weather the loss can be much greater than indicated.

In addition there is no doubt that silage made properly in a sealed clamp will suffer less loss of dry matter than was hitherto thought possible. The assumption was made that the silage produced from the tower should have a concentrate value. This necessitates high dry matter for the crop going into the silo. If, to cut down loss in the field, it receives less wilting, the loss in store, particularly from effluent, will be higher. At Bridgets Experimental Husbandry Farm in 1966, the effect of storing at a lower dry matter content through reduced wilting was examined. By storing at 32·4 per cent DM (10 per cent higher in previous years) the field loss of dry matter was less, but the in-store losses were increased from 3·3 to 15·1 per cent. Either way there is a loss of dry matter. In addition, a lower dry matter silage is less of a concentrate food and this is certainly not justified with the high capital cost involved (see Table 5.3).

Table 5.3
Annual cost per cow of tower silage and clamp silage for 100-cow herd

Tower			Clamp	
		£		£
Tower and silo machinery		5·7	Clamp	3·25
Field machinery		7·5	Field machinery	2·5
	Total	£13·2	Total	£5·75

With grass, the success of the tower system depends on (a) the quality of the crop to be ensilaged, and (b) the amount of loss of dry matter. Quality in this context means percentage of dry matter in the crop together with its digestibility. In many cases it is not easy to obtain a compromise. If, on the other hand, the crop has a high dry matter content when it is cut, one of the disadvantages is overcome. Whole crops—oats and barley—cut in early to mid-July, are becoming popular again for ensilage in towers. When cut, their dry matter content is about 35–40 per cent with an average digestibility of 65 per cent. No wilting is required and losses in the tower should not be any greater than with a well-wilted grass and clover crop. Against this it must be accepted that whole-crop silage will have a lower concentrate equivalent. Maize, if it can be properly established, can be considered in almost the same category, although dry matter content at 30 per cent will be lower. Cut in late September to early October, before the onset of medium frosts, maize will produce up to 25 tons (8 tons DM) per acre of green crop.

It is preferable to divide the costs involved into two categories when comparing, on an annual cost-per-cow basis, high dry matter and sealed clamp silage.

This cost comparison may not be universally accepted, but *on a dry matter basis*, sealed clamp silage is still less expensive than high dry matter silage from the tower. If more sophisticated machinery is used for the clamp, such as full-chop forage harvesters and self-unloading trailers which are required for the tower, there would be little difference between the two systems.

Automatic feeding systems associated with towers are certainly attractive, especially for large dairy herds, although maintenance costs can be high. With more than 500 cows, feeding this way is probably a little slow.

Haymaking

Haymaking techniques in Great Britain have greatly improved in the past 15 years. It is now recognised that there are basically two methods of making hay.

(1) *Quick haymaking.* The crop is cut, immediately conditioned in the field, baled and removed to store in the shortest possible time consistent with its safe-keeping; this means at a moisture content of about 25 per cent. In most seasons, so long as not too much is cut at one time, good hay from a heavy initial crop can be made in up to 72 hours, although the conditioning treatment will invariably cause some loss of leaf depending on the crop being cured. Taking lucerne as an example, 20–30 per cent of the leaf can be lost during making, and the leaf contains twice as much protein and half as much fibre as there is in the stem.

Leaf shattering is not by any means negligible in grasses. Leaf loss is greatest in the later stages of making hay. In the process of curing a crop from about 80 per cent moisture content when it is cut to 25 per cent when it is fit for baling, it is relatively easy to lose up to 40 per cent of the moisture content; it is the last 15 per cent which requires the skill, and invariably when the crop is moved around to further the drying, however gently, there is bound to be loss of leaf.

(2) *Barn hay-drying.* This method avoids a considerable amount of leaf loss as the crop can be baled at between 35 and 45 per cent moisture content for drying to be completed in the barn. In addition there is much less chance of the nearly cured crop being spoilt by the weather.

On dairy farms, storage drying is the most popular way of drying hay in the barn. After drying, the hay remains in the barn until required for feeding. If a barn is suitable for subsequent self- or easy-feeding of the hay, this drying method will be even more suitable, although loose, unbaled hay will have to be used. To date, barn drying of loose hay has not proved to be popular, largely because of high costs. Storage drying has a further advantage over batch drying in that heat is not necessary to achieve a regular turn round for the next batch.

There are farmers now using a variation of storage drying—the Radial-Drying (Dutch) System. This aims to keep costs low by avoiding the need for constructing a plenum chamber and air-tight walls. Baled hay is stacked round a central drying flue, which is enclosed by a "bung unit" at the top. Thus the crop is dried from the centre of the stack outwards.

Tunnel drying is a modification of batch drying. It has the advantage that as the bales can be built up in the field into a stack through which air is blown, it is a cheaper method of drying, although not so efficient.

There is little difference in production costs when comparing quick haymaking and storage drying in a barn. Under ideal conditions for both, the former will cost about £7·50 per ton, including crop production costs, and barn drying will be up to £2 per ton more (batch and tunnel drying is much more expensive). But the wrong sort of weather when the crop is on the ground will increase the cost of quick haymaking out of all proportion, besides very often finishing up with an inferior product. Whilst it may appear to be more expensive, storage-dried hay will produce extra nutrients per acre—these could be equivalent to up to 4 cwt good hay which might be worth £4 more per acre.

Hay preservatives

In spite of these savings, barn drying as a technique is not receiving the attention that it deserves. This is probably because it needs some skill and experience to be assured of obtaining consistently good hay, and also there can perhaps be a limitation in output. Hay preservatives are a possible compromise. A chemical (which acts as a mould inhibitor) based on propionic and formic acid is sprayed into the baling chamber to allow the hay to be baled at a somewhat higher moisture content. This should reduce the loss of leaf in the field.

Tripoding of hay is still seen in certain areas of the north and north-western parts of the British Isles.

Green crop drying

As Table 5·4 shows, when a crop is dried artificially there is less loss of nutrients compared with other conservation methods.

Table 5.4
Grass conservation

	Percentage loss of nutrients		
	Starch equivalent	Protein	Carotene
Hay	45·0	32·5	32
Silage	22·5	10·0	5
Dried grass	5·0	5·0	15

In Great Britain, encouraged to some extent in England and Wales by the Milk Marketing Board, green crop drying was most popular in the years 1947–53, when about 1,000 driers were in use. With feedingstuffs rationed, dried grass was fed to cattle and sheep, but in 1953 when restrictions were lifted on feedingstuffs, dried grass lost its premium value and most of the dried material was sold to the stock food manufacturers for incorporation, chiefly because of its carotene/xanthophyll content, in pig and poultry rations. It was only the most efficient and generally the larger plants which continued to operate, the returns obtained making them better able to compete. Until 1965 there were about 100 of these plants being operated in the country, but since then there has been an increasing revival of interest in green crop drying, due, in the main, to the following reasons:

(1) The high-temperature (up to 1100°C) driers now available are far more efficient in terms of output, fuel economy and labour compared with the smaller and generally low temperature driers of 15 years ago.

(2) Better crops are now more easily grown.

(3) More is now understood about the digestibility of the green crop when dried. For ruminants it also appears that it is preferable if the dried material is chopped to about 1- to 2-inch lengths and then compacted to produce a wafer of about 2-inch diameter and 1-inch thick. Some farmers prefer a large "jumbo" nut or cob and more investigation needs to be carried out on this processing. What is agreed is that dried grass is less digestible if it is finely milled, because in this condition it passes through the animal too quickly.

(4) On the stockless corn farm, grass as a break crop would have more potential if it could be utilised as dried grass, although on the mixed farms under discussion there will seldom be any problem in finding a suitable alternative to cereals.

(5) The rising costs of alternative feedingstuffs.

It is not suggested that many dairy/arable farms will once more carry a grass drier. Table 5.5 shows that the costs involved will largely prevent this, but there could be a re-development of grass drying on a co-operative basis. Encouragement is given in this direction by Farmers' Machinery Syndicates to which, for approved schemes, loans of up to 90 per cent of the purchase price may be arranged over a 4-year loan period through syndicate credit companies. This is already happening with some of the larger farms in this group. It is essential to see, once a contract is made to cut and dry the crop at a certain period or periods of the year, that the contract is, as far as possible, rigidly adhered to on both sides. Only in this way can the expensive plant be kept going throughout the season drying the right sort of material.

Table 5.5
Grass drying costs

Capital cost—3 ton/hour drier plus ancillary equipment and buildings	£100,000
Annual charge—on buildings and machinery	£20,000
,, .. —for 5,000 ton production (labour, fuel and electricity)	£25,000
,, ,, —equals £9 per ton	
N.B. Crop production costs not considered.	

Little work has yet been done on the feeding of dried grass and how it can compete with alternative feeds for cattle and sheep. Early evidence shows that chopped (not milled) dry grass, prepared in the form of a large cob, could well replace the $3\frac{1}{2}$–4 lb to the gallon of a balanced dairy nut, and it could also be used as the protein supplement with cereals.

Straw balancer system

Increasing the productivity of grassland with better management, including the use of more fertiliser, more efficient grazing techniques and better methods of conservation, are of course the usual ways in which the stocking rate on the farm can be increased. On the all-grass farm, apart from buying in feed, there are no alternatives.

Until recently cereals have been showing a very satisfactory return, and for this reason farmers working the dairy/arable farming system have been prompted to grow more cereals at the expense of the grass acres. To compensate for the loss of grass, particularly for conservation, whilst at the same time maintaining the herd size, straw and barley plus a balancer, which could be non-protein nitrogen, can now be used as a replacement for conserved grass.

In terms of milk yield, there is no difference when feeding this ration compared to conventional feeding, but the cost is higher, especially if bought straw is used, and the gross margin *per cow* could be £10 less. However, the crucial factor is the stocking rate, and if by using this feeding system the stocking rate can be improved by up to $\frac{1}{3}$ acre per cow, the gross margin *per acre* could well be £5 better than with conventional feeding.

An above-average yield of cereals must, these days, help to justify the release of land for cropping. Cereals need not always be the crop. If fixed costs are not affected an extension of the cash root acreage will show a better return. Of course a simple expansion of the dairy herd may well be the answer.

Kale

Kale is still popular for dairy cows and followers on these types of farms, and it is most commonly seen in the south and south-western counties. Depending on the district and variety grown, kale is fed from September to March, and when supplemented with hay it is well-balanced for milk production. Marrow stem kale is still most commonly fed before Christmas, as it yields considerably more digestible dry matter than thousand-head kale, but compared with the latter it is not winter-hardy. Thousand-head kale, being less susceptible to frost, is often grown for feeding after Christmas; but it is not uncommon for both these types, particularly the thousand-head, to be sown with rape as a catch crop on the cultivated stubble of autumn barley in August for grazing off in late October and November.

Maris Kestrel is one of a number of hybrid kales which have been developed by plant breeders and is now being grown. Kestrel is a short, stout-stemmed variety giving yields comparable to, if not better than, the marrow stem types. This combined with a higher digestibility and winter-hardiness is certainly an improvement on the older types. Being shorter-stemmed, Kestrel is more acceptable for grazing *in situ*. This is an important point, because wastage of grazing kale can be very high, probably averaging between 30 and 40 per cent. Potential crop yield has been lost in most years simply because in order to produce lower-growing crops with thin stems for better utilisation, other kales have been sown later.

Grazing, whilst being the cheapest way of utilising the crop, should be limited by soil conditions. Unfortunately too often the crop is grazed on wet, heavy and badly drained soils, as a result of which both soil and stock suffer. The direct drilling of kale into a chemically-killed grass sward does leave a much firmer surface for subsequent grazing. It has the further very important advantage of conserving the maximum amount of moisture at sowing time, and this makes the establishment of a kale crop easier.

In recent years on some farms on heavy land, the forage harvester has been used for cutting and chopping the crop for stock. But chopped kale deteriorates quite rapidly, and after 24 hours it is not very palatable. If it could be cut cheaply, and carried to stock in an unchopped state, there would be less wastage than when kale is grazed.

Other crops

The distribution of arable cash crops in Great Britain has been discussed in Chapter 3. On the dairy/arable farm this pattern is much the same. With barley or oats grown for feeding, wheat is most often the cash crop grown to utilise to the full the existing equipment on the farm.

Potatoes fit in well on many of these farms. They are always a good preparation for wheat, and the chats from the crop are a useful by-product for feeding to dairy followers or even to the herd itself. Where grown, sugar-beet can be included in this category. The tops, fed wilted, are a very valuable food for dairy cows. One acre of sugar-beet tops (average yield 12–14 tons per acre), if they could all be saved, would supply nearly as much starch equivalent per acre as one acre of swedes (20 tons per acre). As silage, tops are also useful provided they can be ensiled in a clean condition. Stock farmers will often grow large-topped varieties with this by-product very much in mind.

Mangolds have declined markedly in popularity in recent years from over 150,000 acres in 1957 to 36,000 acres 10 years later. The reasons for this are not hard to find. It is an expensive crop to grow, and whilst it can lend itself to the same programme of mechanisation as sugar-beet, grown on its own it will not repay the cost of the fairly heavy investment in the machinery involved. Mangolds are grown on dairy farms in the more southerly parts of the country where they are heavier croppers than swedes and turnips.

Like the mangold, and for the same reasons, the acreage of swedes and turnips has also declined; in this case, almost by half in the last 10 years to about 290,000 acres. Swedes and turnips are associated more with sheep and fattening cattle than with the dairy herd, and on dairy farms even in the north and north-west of the country, silage is largely replacing these root crops, although mechanisation of these root crops could halt this trend,

Perhaps another contribution to the falling acreage of root crops for livestock feed is the fact that they are no longer regarded as the cleaning crop on the farm. In this respect they have been replaced by the cereal crop. Late-sown spring barley and/or early-harvested winter barley gives as good an opportunity as root crops for cleaning a field of grass weeds, if not better and it is, of course, cheaper to control annual weeds in the cereal crop than in the root crop.

Farmyard manure

There is on the mixed dairy farm these days a surprisingly changed attitude towards farmyard manure. Its use is no longer taken for granted as had certainly been the case for very many years indeed. Now, on many such farms it is looked upon as a nuisance, something to be got rid of in the easiest and most convenient way! In spite of mechanical handling, farmyard manure is an expensive commodity to move about. Naturally most farmers appreciate its value because of the physical effect it has on the soil, and it is this that makes it impossible to give it a proper cash value.

Because of heavy handling costs, and the fact that on lighter soil it will help to contain moisture, farmyard manure is still used on some cash crops, notably potatoes and sugar-beet, and this is where it fits in so well on the farm. There can be an abundance of straw from the cereal crops for the yarded herd and followers during the winter to produce the farmyard manure for the cash root crop. On average 30 cwt of straw per beast is needed for the winter months in a covered yard, and the 11-cwt dairy cow will make, with the straw, about 6 tons of dung in that period.

Only on a few farms will covered dungsteads still be in use. Whilst undoubtedly moving the dung twice, i.e. from the yard to the dung-stead and thence eventually at a more suitable time of year to the field, will mean a "shorter", better quality dung, heavy labour charges can scarcely justify this these days. Usually the dung is left in the yards, and this is a good practical way of conserving liquid manure until late autumn, when it is carted straight to the field for, it is hoped, immediate ploughing-in.

It costs about £3 to bale and carry an acre of straw (approximately 15 cwt) which then has to be stored. This, and the unfavourable attitude now held by many towards farmyard manure, together with the fact that it can be more profitable to sell straw to grass farms in the west, has led many farmers on the mixed dairy farm to abandon the traditional use of straw for bedding, and to adopt the cubicle system for housing cattle (see Chapter 7). This in turn has led to the problem of slurry disposal.

Slurry

By the various Rivers (Prevention of Pollution) Acts introduced by the British Government, it is obligatory for all slurry to remain on the farm. A cow produces between 15 and 20 gal per day of slurry i.e. dung, urine and waterwash. Thus for a 100 cow herd, 2,000 gal has to be dealt with daily. And this in turn means that some form of storage for the slurry is normally necessary. There are many systems of handling slurry, none of which is entirely satisfactory. With cubicles, and where the slurry can be scraped out daily to a dung-spreader or storage compound, the initial cost will be about £40 per cow. At double the cost, with cubicles plus slats, the slurry can be stored in an under-floor tank to be disposed of eventually through organic irrigation or by way of a trailer sludge tank.

A slurry-storage compound, costing approximately £10 per cow, is one of the more satisfactory ways of holding the slurry during the winter months. Additionally, on many dairy farms a surrounding compound of sleepers supports a straw-bale wall. During an average summer, the slurry will dry out reasonably well for handling in a semi-solid state in the autumn.

In terms of plant food, slurry is of little value, although how much it contains depends on the degree of dilution. From cows, 1,000 gal of undiluted slurry contain about 40 units each of nitrogen and potash, and 10 units of phosphate. When mixed with rain or washwater, it can be appreciated what little significance slurry has when applied to the land. The general tendency on many of these farms is, during the winter months, when there are no suitable facilities for long-term storage of the slurry, to apply it (with a trailer sludge tank or the like, or possibly by organic irrigation) to the most convenient field, which is very often in stubble prior to late ploughing.

Disposal of slurry in this way has not been practised long enough to see what long-term effect it has on the soil. Well-drained soils are naturally the most suitable, but very light sandy soils may be too free-draining. And of course during winter especially, to avoid waterlogging, the less water used the better. Lime is used up as organic matter decomposes, and so the lime status of the soil must be watched when large quantities of slurry are applied. Probably a total of 15,000 gal (in three applications) per acre of diluted slurry should not normally be exceeded.

The growing difficulties of slurry disposal over the past few years have at least made farmers realise that this is a problem, and certainly when new buildings for livestock are being planned it should receive prime consideration.

DAIRYING AND FATTENING FARMS WITH ARABLE CROPS

Farms in this category include not only arable and dairy, but in addition one or more other livestock enterprises.

Fattening cattle

Intensive cereal feeding to beef has pointed the way to quicker and more profitable ways of fattening cattle. The profitability of these enterprises has depended in great part upon the relative prices of bought-in calves and barley. Barley beef from cattle of 1 year of age has its limitations, but there are farmers who still successfully fatten beef in this way using Friesian steers from the dairy herd. Grass of course is not used for barley beef, but it has a very important part to play with *semi-intensive* 12–18 *month old beef*. Two main systems can be considered in this way.

(i) Grass can be brought into an intensive system using the autumn-born calf. This system has evolved because many intensive cereal feeding systems are no longer very profitable, and the meat produced is considered by many to be not so acceptable. Consequently the fattening period is extended by introducing grass, and finishing off the animals at anything from 12–15 months of age at about 8·5 cwt liveweight. The method is simply to *interpose* a summer at grass *between* the calf winter and the period of high-barley or other concentrate feeding. The animal reaches about 8·5 cwt in January when there should be a better price obtainable. Although efficiency of food utilisation is less than with barley beef, it should be remembered that some part of the liveweight gain is made relatively cheaply off grass, and with a seasonal rise in beef prices, farmers find that this is a system well worth considering.

(ii) A less intensive method of fattening off grass is to bring silage into the final fattening period. This again is with autumn-born calves reared inside during their first winter, and perhaps including in their ration a small amount of really good silage. The summer is spent at grass, and for the second winter, the stock are yarded and fed good quality silage and a small amount of hay and concentrates. They are sold off from the yards at 15–18 months old weighing 9–9·5 cwt. Very impressive gross margins of up to £50 per acre have been recorded with this system, making it more than comparable to cereal growing and to dairy farming at traditional rates of stocking.

Single-suckled autumn-born calves can also be fattened in this way. The cows and calves are yarded on self-feed silage in the winter, but with the calves getting creep feed, depending on the quality of the silage. It is normally possible to turn out a good strong calf in the spring, capable of dealing with the grass, as well as the grass flush of milk from its dam. Both may be paddock-grazed together, allowing ½ acre of grass per cow and calf over the grazing season. The calf is weaned in summer to autumn, for finishing off in the spring on silage, some hay and concentrates.

The systems described are all concerned with autumn-born calves, and attempts have been made to modify the methods of fattening to suit spring-born calves. They are reared on grass having been weaned from cold-milk feeding; the winter is spent on good quality silage, with some hay and concentrates, and in their second summer they are finished off under an intensive paddock-grazing system for selling at about 9 cwt liveweight in the autumn.

The Hereford/Friesian cross steer is seen at its best with semi-intensive beef, and some farmers are having some success with the Charolais on dairy breeds.*

* Current high prices for the Hereford/Fresian calf may make more dairy bred calves an economic proposition.

With all these systems, paddock grazing is an advantage, coupled with the liberal use of nitrogen—up to 300 units per acre in the season. The standard of grazing is of supreme importance to maximise liveweight gain from the relatively cheap grass; 1·7–2 lb liveweight increase per day can be achieved with an *average* stocking rate throughout the *whole* season (April to late October) of 3 beasts per acre.

No less important than the grazing is the quality of silage made. High dry matter (25–30 per cent) silage should be the aim. When the cattle are yarded, liveweight gains of over 2 lb per day can be achieved from self-feed silage, 2–3 lb hay and 6–7 lb barley.

Grass sheep

When dairying and arable are combined with livestock farming, in many cases the grass sheep flock can fit in extremely well. Fat lamb production is normally the main enterprise, and although at certain times during spring and early summer there may be competition between the herd and sheep flock, more intensive grazing techniques for both the herd and the fattening lambs—see Chapter 7—has brought about an improvement of stocking rate with what should be a consequent lessening of competition. At other periods of the year, notably in autumn and early winter, doubtless many dairy farmers welcome the presence of the flock because it is then that sheep can fulfil their natural role as scavengers.

Apart from grazing stubbles, the flock can do a tremendous amount of good on cow pastures simply by eating down the surplus so that the grass is in good close-knit condition, and better able to stand up to winter-kill. This is especially the case with ryegrass leys. With timothy, meadow fescue and cocksfoot, foggage production for dairy followers and beef cattle may be practised if conditions allow.

Store lambs can be as profitable as fat lamb production especially if, because of a late spring, concentrates have been fed to the latter. And like the scavenger ewe, store lambs can be most useful in the autumn.

Pigs and poultry

Pig keeping is now very intensive, with many systems being run completely indoors. But on a large number of farms where pigs are kept, litters are still reared outdoors prior to indoor fattening. The last year of a ley may be utilised in this way. As a secondary enterprise, the sale of really healthy weaners at 8 weeks of age can sometimes show a very impressive gross margin per acre in addition to the increase of fertility afforded to land so treated.

Much the same can be said of *poultry* arks moved daily across a field. This is a practice now very much in decline. It was common to house laying birds this way; a standard size unit holding 25 birds and going over the ground once in the year would supply per acre the equivalent of about 90 units of nitrogen, 120 units of phosphate, and 45 units of potash, in a readily available form and with a minimum of loss. However, labour demands are now too high, and although there are some farmers rearing pullets in fold units (about 40 per ark) there can be no control over light hours, which is accepted as being essential for subsequent egg production.

On some farms the battery hen fits in extremely well with other livestock and indeed with arable cropping. The poultry manure from the battery unit can be put to profitable use. If for example a 30,000-bird laying unit is being operated, in terms of fertiliser, the poultry manure is worth something over £2,000 per year. Of course none of the plant food is as readily available as from the fertiliser bag, and there is, in particular, quite a loss of nitrogen before it is actually applied. Nevertheless if precautions are taken to minimise this loss, a very valuable manure for crops and grass, bearing in mind an imbalance of potash, can make a sizeable saving in the use of chemical fertilisers.

FATTENING AND ARABLE FARMS

Beef fattening

Where conditions are unfavourable for dairying, cattle and sheep fattening take on the role of major stock enterprises in conjunction with arable farming.

The semi-intensive beef system already outlined is increasingly practised on farms in this group, and in addition 18- to 21-month beef is largely being produced from arable by-products instead of grass silage. This is chiefly associated with the arable areas of the eastern counties of England and Scotland, although it is used in other districts where there are plenty of by-products.

Pea-haulm silage, with a crude protein content of up to 17 per cent, wilted sugar-beet tops and beet top silage can form a valuable part of the ration, as also can beet pulp, either wet or dry. All these foods may be used in the animal's second winter fattening stage. Much the same can be said of potatoes. As standards of potato grading rise and the trend towards pre-packing continues, more surplus potatoes are becoming available for cattle feeding. As long as they are reasonably clean and disease-free and are introduced gradually into the diet, they are very satisfactory.

Traditional yard-fattening is still practised using fairly mature store cattle weighing between 7 and 9 cwt. Turnips used to be the mainstay of the diet, and from 120 lb up to as much as 200 lb each were commonly fed, plus about 10 lb hay and 4 lb concentrates. Such a ration was capable of putting on between $1\frac{1}{2}$ and 2 lb liveweight gain each day. But it was accepted as expensive meat with the compensation of good farmyard manure for the root crops. This very traditional system is now being replaced, with completely satisfactory results, using silage instead of turnips. However it is suggested that improved husbandry and feeding techniques could create a revived interest in the turnip for cattle feeding.

Summer fattening of beef cattle is also a major feature on many of these farms. The basis of this system is that the cattle are bought as 18-month stores and fattened off at $2-2\frac{1}{2}$ years of age.

Many of the best summer fattening pastures are permanent grass, although not all permanent pasture is suitable for fattening cattle. It is true to say that on many farms there are good and indifferent pastures, with the former receiving most attention and much more of the fertiliser! Even at their best, stocking rates with this system seldom exceed one to the acre, and because of this more attention has recently been paid to fattening on leys, at one time thought unsuitable.

This has not come too soon, because it is difficult to understand why permanent pasture has been considered so superior for this system of fattening. Investigations nearly 30 years ago showed that only about one-third of the permanent grass in England and Wales was capable of fattening one beast to the acre during the grazing season, compared with the ley which was capable of a considerably higher performance. And since then the production from leys has improved considerably more than it has from permanent pasture.

In contrast to the dairy/beef crosses used in the systems previously described, most dairy stock with the exception of the Friesian have been found to be unsatisfactory. The Hereford, Devon, Sussex, Lincoln Red, Beef Shorthorn and Aberdeen-Angus and crosses of these breeds make the best liveweight gains under this method of fattening. But breed is not everything, and many who fatten successfully on grassland find that it pays to buy some of their cattle in the autumn to ensure proper feeding and handling prior to summer fattening. Another advantage can be the high cost of store cattle in the spring compared to autumn prices, and in some years this will more than offset the cost of wintering.

When pasture is being stocked up in the early spring, the usual rate is one beast per acre. Normally once assigned to a pasture, stock stay there until they are fat in late June to early July, having put on a liveweight gain of about 2 lb per day. It is seldom that they are moved to a fresh field during fattening, as this could severely check their rate of progress. A second batch of animals immediately follows to be ready fat in the autumn, and very often to enable them to be finished at that time some cereals are fed in addition.

It is easier to fatten on permanent pasture than on leys, and with these latter in particular, a form of paddock grazing and indeed strip-grazing is now being tried. Heavier stocking rates are possible, but it is contended by many that there is a slower performance per animal. Others would argue against this traditional view. Quite impressive returns per acre can be achieved with both these more intensive grazing systems, although they demand a higher plane of management. With a better price incentive for beef, it is possible to see an extension of these more intensive grassland fattening systems.

Arable sheep

There are still farms in this group which maintain an arable sheep flock. Down sheep are the main breeds involved, and with some specialist breeders, a valuable product from the system is breeding rams to be sold for use in grass sheep flocks to sire lambs for fat lamb production.

During summer, the flock is fed on leys; it is only folded on arable crops and by-products in the autumn and winter. Sugar beet tops, rape and kale are chiefly utilised, whilst root crops such as swedes and turnips, because of their high labour demands, are declining. This is particularly the case in the south; but in favoured growing areas, and with more mechanisation, swedes and turnips are still an important livestock crop. To fence in the stock, hurdles are still used, but more and more farmers are turning to the close-mesh electric fence.

In addition to rams, farms on an arable sheep flock system may produce early fat lambs, and also, later in the year, can fatten bought-in or home-produced store lambs.

Further reading

The student is referred particularly to the Rothamsted Ley-Arable Experiment, which is dealt with in the following publications:

Report of the Rothamsted Experimental Station (1961)
Rothamsted Experimental Station Field Experiments and Work of the Departments (1965)
Report from High Mowthorpe Experimental Husbandry Farm

Chapter 6

LIVESTOCK PRODUCTION ON HILL AND UPLAND FARMS

Mainly because of topography, many hill and mountain areas of the British Isles are such that the land cannot be cultivated. They have therefore been left down to grass. In some cases the quality of the grassland has been improved in the course of centuries.

These areas play a great part in British agriculture.

No attempt will be made in this chapter to describe the hill and upland farms in detail, but rather to describe their role in relation to the lowland farm. Traditionally the hill country has supplied store and breeding animals for the lowland farmer to fatten, or to use for breeding within the cross breeding system which has been mentioned previously. Sheep have always taken precedence over cattle on our hills, and the breeding of cross-bred ewes is one of the major features of our hill and upland farms.

HILL AND UPLAND SHEEP FARMING

The main hill breeds of sheep are known by the names of the areas in which they originated. Scotland has two predominant breeds of hill sheep: the Blackface and the Cheviot. The Scottish Blackface is found, in the main, on what is known as "blackland" where the natural vegetation is mostly heather. The breed contains several different strains, each tending to have its own characteristics of size, thickness of fleece, length of wool staple and so on. To name two: one often hears of the Newton Stewart and Lanark types; but the differences are such that the ordinary person has difficulty in distinguishing between them. Perhaps the most important characteristic of the breed as a whole is its ability to forage, and to survive in severe weather.

The Cheviot is located mainly on the grassy hills of Scotland and there are two main strains of the breed. The South Country Cheviot type is a true hill breed, and is found particularly in the area of the Cheviot Hills, where the breed had its origin, and where the vegetation is of the nardus/fescue type. At the beginning of the 19th century Cheviots were introduced to the county of Caithness in the north of Scotland, where they became established on semi-arable land and from where they spread to other parts of Scotland. In time a larger and more prolific type of sheep emerged, which became known as the North Country Cheviot, and this remains an important breed in Scotland today.

Moving from Scotland to the four Northern English counties and Yorkshire, different hill breeds are found. Apart from the Lake District the area is virtually the Pennine Range. The basic hill breed of the Pennines is the Swaledale, which is not unlike the Scottish Blackface, but in fleece characteristics and in face markings is quite distinct. Generally Swaledale ewes are hardier than Blackfaces and have not been subjected to quite the same "improvement". They are much valued in the very exposed bleak conditions of the Durham and Yorkshire moors.

As well as the Swaledale there are other hill breeds, namely the Rough Fell found in Cumberland and Westmorland, and Lonk in Lancashire and the Derbyshire Gritstone in Derbyshire. A quick glance at these three breeds reveals their basic Scottish Blackface character, but they are not very important except in their home locations. In the Lake District in the high hills around Keswick one other breed, The Herdwick, is found. This is a small breed said to be of Spanish origin. It is brick-red in colour and very late maturing.

These breeds together with the Welsh Mountain are the main hill breeds of the British Isles. The main characteristics of all hill breeds is that they are hardy, tough, good mothers with an adequate milk yield. Another good characteristic is that they are very good graziers.

The shepherd's year

On hill and mountain farms the system of sheep farming is fairly standard. The ewe flock, comprising regular age groups, stays permanently on the farm, except for the ewe hogs which may be away-wintered on low ground during their first winter, and in this way the flock becomes acclimatised to the terrain on which it has to forage. When hill farms change ownership, the new owner generally buys the sheep stock at valuation and in addition pays a price per head which is a monetary recognition of acclimatisation.

Basically the shepherd's year is concerned with the grazing needs of his flock. Each day he, accompanied by his well-trained dogs, walks, rides, or enjoys the comfort of a motor vehicle round the various parts of the hill to look after his sheep, checking on the grazing needs of the flock, and encouraging them to graze appropriate parts of the hill at certain times of the year. Knowledge of the grazing potential of different parts of the hill is essential to the hill shepherd.

Seasonal duties are very similar to those of the lowland shepherd. Lambing usually begins in April, but on a really high farm it might be delayed until May. This is because of the great advantage to the young suckling lambs and their mothers of a bite of fresh spring heather or grass, to see them through this critical stage. The ewes are generally brought down to the "inbye" fields surrounding the house and buildings for lambing. Lambing takes place in the open fields, with natural contours and walls or dykes providing all the shelter the ewes and lambs require. Occasionally, weak lambs or ailing ewes are penned in shelter to allow for special treatment.

Within a very short time of birth the lamb is up and away with its mother to the hill, and at once it can be faced with hazards. The weather at lambing time plays a decisive role, particularly because losses in lambs can be high if conditions are not reasonably dry. Late storms and lashing rain are much more severe on young lambs than cold but dry conditions. The ideal is warm and dry weather. Mountain streams in flood can result in many lambs being drowned in trying to follow their mothers, and natural hazards such as foxes and carrion crows may constitute quite a danger.

Under such tough conditions the ewe has her work cut out to cope with a single lamb, and in fact twinning can be an embarrassment if there is no low-ground grazing available to see the ewes and lambs through the weeks immediately following lambing. A farmer is therefore generally satisfied with a 100 per cent lamb crop from a true hill flock, but in practice often ends up with less.

Some six weeks after lambing the flock is gathered from the hill for castration and tailing. The latter operation is not always performed as in some areas it is common practice for hill ewe lambs especially to retain their tails to provide warmth and shelter in storms to come. After these routine tasks the flock is turned away to the higher parts of the hill and left to graze until shearing time in late June or early July. In many cases the gathering and the clipping is carried out on a communal basis—neighbour helping neighbour. At the time of clipping, ear marking or horn branding is carried out and a thorough count of the flock is made.

The ewes and lambs are then returned to the hill where they remain until dipping some weeks later. Dipping to control insect pests is of great importance on hill farms, because many diseases of hill sheep are transmitted by insects. Ticks, sheep keds, lice and maggots of the bluebottle fly are examples of insect pests which must be controlled by dipping. Special facilities in the form of pens and a runway leading to a dip bath must be provided. Appropriate dipping material in a given quantity is added to water in the dip bath and each animal is completely immersed in the solution for a minute or so. Very often there is a shallow foot bath incorporated in the dipping layout through which the sheep are run, and in which there is an appropriate solution to deal with diseases like foot rot which can lead to lameness and loss of condition.

After dipping the flock is again returned to the hill and remains there until August/September when the sheep sales commence. After weaning, the flock is sorted into lots; the wether lambs for sale, the ewe lambs for future breeding and sale, the ewes to be kept, and the draft ewes for sale.

Rams, when not running with the ewes, are given special attention in the "inbye" grazing fields, and generally receive some supplementary feed to keep them in good condition.

The sales

The hill farmer has little else to sell other than lambs and wool. The lambs are mainly sold on a store market where the price is subject to the laws of supply and demand, unlike his lowland counterpart who is selling a finished animal for which there is at present a guaranteed price. During the autumn months, as well as dealing with this work at home, the hill shepherd spends much of his time at markets, for the wisdom gained from the marketing of his stock plays an important part in the overall prosperity of his farm.

Many hill wether lambs are sold for fattening on arable crops in the marginal areas where foggage, turnips, swedes and rape are important crops. The overall trade for fat lambs, coupled with the availability of sheep feed, basically determine the price. If store lambs are wanted then the hill farmer can get good prices. Draft ewes are the other main store item he has to sell—and here again their price is reflected in the overall profitability of lowland sheep farming. Some years trade is well up, and in other years well down.

Finally there is the wool to sell. In this case there is at present a guaranteed price. Clips can vary in weight from 2 lb to around 4 lb per animal, and variations can also occur between season and season. The type of wool varies from breed to breed, not only in length of staple, but also in fineness, and in the percentage of hair in the fleece. The result is that wool from a particular breed is often sought for special purposes. Blackfaced wool, for example, is well suited to carpet making, the filling of mattresses, and for the hard wearing Harris type of tweed cloth.

In recent years there have been various suggestions as to how profitability on the hill sheep farm can be improved. One method tried has been the fattening of lambs on a cereal/protein mixture, with the idea of becoming independent of the fluctuating store market. Unfortunately the food conversion ability of hill lambs is not particularly good, and whilst some forward-thinking farmers were keen at the outset the practice has not grown, largely because the feed, labour and in some cases housing costs have made the system no more attractive financially to date than selling lambs as stores. Some hill farms in suitable areas are able to grow rape, swedes or turnips on their "inbye" land and may fatten their own lambs on conventional homegrown foods.

The overall prosperity of lowland sheep farming reflects back up the hill. When fat lamb production is profitable there is a good demand for cross-bred ewes, and therefore for the draft hill ewes out of which the cross-bred ewes are produced.

Breeding

Owing to adverse winter conditions the majority of hill ewes only remain in the flock for three or four lambings. After their last crop of lambs has been weaned, these ewes are sorted and kept apart until they have improved in condition somewhat, after which they are sold as "draft ewes". These so-called draft ewes are purchased by the upland or marginal land farmer who will keep them under much better conditions than the true hill farmer, thus enabling these ewes to have another one or two lamb crops. A "longwool" breed of ram is used on them to produce a cross-bred lamb, the ewe lambs being sold for further breeding on lowland farms, and the wether lambs either sold for fattening or finished by the farmer himself. The most important are the ewe lambs, as it is these which grow on to provide the grassland breeding ewe, so well known on lowland farms as the basic ewe for fat lamb production.

Both Cheviot and Blackface draft ewes are mated to the same "longwool" breed of ram—the Border Leicester. Sheep breeders in the

Borders pride themselves on their ability to breed top-class pedigree Border Leicesters, and a visit to the Kelso Ram Sale in September bears this out. Although this is a principal sale for Border Leicester rams, rams of most breeds are offered for sale, and to the person interested in sheep a visit to Kelso is most rewarding.

The Border Leicester mated to the Cheviot, produces the Scottish Half-Bred and with the Scottish Blackface ewe produces the Mule or Greyface.

Of all the cross-bred ewes available the most important is undoubtedly the Scottish Half-Bred and some would go so far as to say this is the best cross-bred ewe available to the lowland farmer. It is much sought after and pens of top quality Scottish Half-Breds make high prices at the autumn breeding sales. The price can be so high that an alternative cross is often sought, and here the Mule or Greyface is gaining in popularity. This particular cross, although very prolific, has never been as popular, although it has made progress over the last few years.

There are slight variations to this basic pattern of sheep farming in parts such as the Borders. There is a strain within the Border Leicester, known as the Bluefaced or Hexham Leicester—this is a smaller, lighter-framed sheep than the true Border Leicester, and has a characteristic blue skin. It is claimed that rams of this type when crossed with the Scottish Blackface give a better Mule, compared with the ordinary Border Leicester. Reference has already been made to "better ewes"; how is this measured? Rightly or wrongly "quality" in breeding ewes is assessed in terms of conformation associated with what has come to be known as "stylish" face markings and general appearance. Border shepherds are past masters in bringing out the sheep for sale. As the Perth Bull Sales are to the beef breeder, so are the Border sales, and perhaps Kelso in particular, to the sheep breeder. At both Perth and Kelso one sees the best of the respective breeds being presented for sale in superlative condition for which Scots stockmen are famous.

New developments

However, even in traditional sheep strongholds ideas are beginning to change. The supremacy of the Border Leicester as the only crossing ram for the Scottish hill breeds is being questioned. The Colbred ram, now a pure breed, but originally from a fourway cross, has been used as an alternative. The results of using these particular rams on the major hill breeds have proved rather disappointing, and so far they have **not** reduced the popularity of the Border Leicester. Rams like the "Improver", again another cross-bred in origin, are being used, and great interest is also being shown in the use of the Finnish Landrace ram. The aim behind the use of such rams is to try to obtain a cross-bred ewe which is more prolific, and has a greater milk yield than the traditional Border Leicester crosses. The hill farmer is aware of the importance of these new rams, but so far, however, is confident that the Border Leicester cross is still the one required, and he is in the main continuing to produce these.

Another important relevant factor determining the breed of ram used is of course the value of the wether lamb. It has already been mentioned that these wether lambs are sold, as stores, for fattening on fodder crops. The feeders know what they are dealing with in terms of Border Leicester crosses, but are less sure when it comes to buying lambs sired by unfamiliar breeds. In particular the lambs sired by Finnish Landrace rams seem very much "on the leg" and consequently do not give the appearance of an early finish—as do the Border Leicester crosses.

The north of England

After drafting from the high moors, the north of England breeds are crossed with the Wensleydale ram, producing what is commonly termed the Masham; the many market towns of these areas, for example, Kendal, Hawes and Skipton are centres where these ewes can be obtained at the autumn breeding ewe sales.

As with the Border Leicester the Wensleydale too has strains within the breed, and except to the true "dalesman" these tend to be a little confusing. Basically the main strain is termed the Teeswater and these rams, for example, mate better than the true Wensleydale with the Rough Fell. Whichever ram is used, however, the general term Masham is given to the progeny. Masham-type ewes are much sought after by lowland sheep farmers as they are very prolific, good mothering and milking ewes.

SUMMARY OF THE SYSTEM

To summarise, the hills and moors of Scotland and Northern England and Wales provide a "reservoir" of tough, hardy, good milking, good mothering breeds of sheep. These, having bred three or four crops of lambs in the high hills and exposed moors, are then sold to a farmer "lower down the hill" who crosses them with either a Border Leicester or Wensleydale ram to sell to the lowland fat lamb producer an excellent type of cross-bred ewe having all the hybrid vigour described in a subsequent chapter.

There are however intermediate situations where a farmer farms both high land and marginal land, and when the two operations of pure breeding and cross-breeding are done on the same farm. Some farmers keep cross-bred ewes, and a very popular cross on many of the poorer, high moorland farms is the Blackface x Swaledale.

Only within recent years has this system of cross breeding on upland farms been practised to a commercial extent in Wales. The Welsh Half-Bred, the cross between the Welsh mountain ewe and the Border Leicester, has become a very popular breeding ewe, and particularly so on the sheep farms of the West. The West Country farmer finds it easier to obtain this ewe than going to the North or Scotland for his breeding stock replacements, and sheep sales are now becoming well-established in market towns such as Kington and Knighton.

ECONOMIC PROBLEMS

Whilst the foregoing outlines the general pattern of sheep production on hill farms, the economic difficulties which have been with the sheep industry as a whole through the 1960's, have led to re-thinking and considerable changes, largely on hill farms with an acreage of lower "inbye" land available for the production of good grass and in some cases small acreages of cereals and roots. The underlying aim now is to produce more lambs from the same number of ewes, using the "inbye" land to provide better nutrition for the ewes throughout the year. It is considered by many that the traditional survival-of-the-fittest, giving a low lambing percentage system, is quite uneconomical; quite a number of upland farmers now look for anything from 120–150 per cent to 150 per cent lambing, even from traditional hill breeds like the Scottish Blackface. This may be achieved by better nutrition, especially during the weeks preceding tupping time, and the feeding of concentrates some six weeks before lambing, and if necessary after lambing, to keep the ewes in milk. In this way the ewe is able to cope with more than a single lamb, and the number of lambs coming off the hill from the same number of ewes is of course increased significantly.

Since herding costs represent a considerable proportion of the year's outlays, management is being adjusted so that the number of ewes in the care of a full-time shepherd steadily increases. From the once traditional 500 ewes or thereabouts, the number on some farms has reached double or more this figure.

CATTLE

Many upland farms now run a beef breeding cattle enterprise, generally with a herd of cross cows producing single suckled calves for sale at the calf sales which take place at many centres from October onwards. On favoured farms where adequate winter fodder can be grown, calving starts in November, and in some cases all calves are born before the end of the year. If conditions are more severe, spring calving may be preferred to give the cow and young calf the advantage of early spring grass. The fact that attractive subsidies have been available for cows and calves bred on upland hill farms has encouraged a very significant increase in the breeding beef herd over the past two decades, and on some upland farms the production of beef calves now takes precedence over sheep.

This cattle development has led to great attention being paid to the problem of winter food for the cows. Hay or silage, or both, are fed and the techniques of making these have advanced considerably.

On some farms the cows are outwintered, and whilst there are advantages in this method, most farmers are now coming round to the idea of some type of housing for the winter months. One of the main problems of outwintering is the poaching of grass fields which may have been improved and which are needed both for grazing and for the production of hay or silage. Another difficulty may be in getting the daily ration of feed out to the fields in really wet weather or when there is a heavy fall of snow. Whilst hill cows are predominantly cross cows, the bulls used are from pure breeds of beef cattle. Aberdeen-Angus, Hereford and Beef Shorthorn are the most widely used breeds, but in recent years a number of other beef breeds have been tried. Whatever the breed of bull used or the type of cow, the aim is to produce a growthy calf which will flesh quickly in the hands of the feeder on the low ground farm.

This reservoir of beef calves has now become an integral part of livestock production in Great Britain, and its fortunes, like those of hill and upland sheep, very much depend on the prosperity of the low ground farmers who buy the store stock.

CONCLUSION

How the stock, both cattle and sheep, having been bred on hill and upland farms, is sold to lowland farmers for intensive livestock production, is described in the next Chapter.

Chapter 7

INTENSIVE LIVESTOCK SYSTEMS

In all spheres of business, units are becoming larger, and in the United Kingdom agriculture is no different from any other industry. In general the changes have not been organised, though the British Government have provided grants for the amalgamation of farms, for the removal of ditches and hedges to enlarge fields, and for the new buildings arising from such amalgamations and reorganisations.

Grants have also been available for the erection of up-to-date large new buildings to meet farmers' requirements for larger specialised units, such as piggeries or milking parlours. In other instances general-purpose buildings have been erected to provide a roof and sides, into which fittings have been put to make pens for any or all of the following: (a) calves, (b) store cattle, (c) fattening cattle, (d) suckling calves, (e) lambing ewes and (f) weaned pigs, for permanent housing or temporarily to liberate other buildings for disinfection. Such general-purpose buildings may be used, at certain times of the year, to store potatoes for several months in winter and corn for a few weeks after harvest. If the floor is of concrete with water laid on to strategic points, the general-purpose building can be easily prepared for livestock. It has been British Government policy to subsidise the erection of such utility buildings.

In 1964 a Committee was set up by the Minister of Agriculture under the Chairmanship of Professor Brambell to discover whether the modern systems of keeping large numbers of stock and birds in confined spaces was cruel and produced hardship and discomfort to the stock. Arising from their report (1965), the Ministry of Agriculture has recommended:

(1) Acceptance, subject to safeguards, of the practices of stalling and tethering cattle.

(2) Acceptance, subject to safeguards, of totally slatted floors for cattle and pigs.

(3) Acceptance of sow stalls and cubicles with tethering and without straw.

(4) Revised and reduced space allowances for intensively-kept poultry.

(5) Acceptance for poultry, subject to safeguards, of beak trimming, the use of spectacles, and dubbing in certain circumstances.

More attention has been paid to domestic fowls and turkeys than to cattle and pigs because the former are more frequently kept on factory lines. Whether these recommendations will become the last word on this debatable subject is open to question, but the suggestions may have a profound effect upon the whole question of intensive livestock production, and some of the schemes, as described later, may have to be modified as a result of new legislation.

Intensive units of livestock have been developed in two main ways, namely, (a) where the enterprise is quite separate from a farm, and (b) where it is an integral part of the farm. These two different sets of circumstances have their own problems.

INTENSIVE LIVESTOCK UNITS
SEPARATE FROM THE LAND

Poultry

Such intensive units were first developed with poultry where derelict factories were converted into intensive poultry houses. All that was needed basically for such development was a good floor, not very much head room (8–10 ft), the possibility of (i) plenty of ventilation (this may be completely controlled by fans) to control humidity and temperature, (ii) pens varied in sizes to suit the various birds housed, (iii) a good supply of reliable drinking water, (iv) a supply of electricity for power and to control light which influences egg production, (v) what may be most important of all, some certain way of disposal of droppings or dung, and (vi) a building that (when empty of stock) may be easily and completely disinfected. To meet this last need, the building is usually constructed in sections, so that periodically one section can be sealed off from the next, emptied of stock and then completely disinfected. The building must also have a good access road so that (a) foods can be delivered in bulk into large hoppers, (b) produce can be moved away easily, and (c) manure, or droppings, can be dispatched for direct use, or dried and used as feed for cattle or sheep or as a manure in horticulture.

With most intensive systems birds never go out-of-doors, as is seen from the following progressions:

Broilers

(1) In incubator (temperature 40°C);
(2) In deep litter houses in batches of the same age, where birds remain until they are slaughtered as broilers at 8–10 weeks of age, weighing about 3 lb.

Layers

(1) In incubator (temperature 40°C);
(2) In brooder house kept in batches according to heat required (15°C at six weeks);
(3) From the brooder house into deep litter house where birds remain until they are ready to be moved to their laying quarters;
(4) Some, on the point of lay, remain in deep litter houses where they may be kept in units of 1,000 laying birds. Others will go into laying batteries for single birds, or for several birds housed together, in a pen.

The question of size of battery cages and number of birds per cage is being debated as a result of the Brambell Report. It is accepted that the battery cages must, for economic reasons, be several tiers high (the exact number is also under discussion). The alleyways between the rows of birds muct be sufficiently wide to facilitate the passage of staff for supervision, feeding, inspection of birds, and also for the collection of eggs.

With deep litter systems, the removal and disposal of the droppings is now a very great problem; this is also very acute with battery cages. Some have found that it has paid to dry and sell the manure to small gardeners in cwt, $\frac{1}{2}$-cwt and $\frac{1}{4}$-cwt lots. More recently it has been sold for feeding as a protein supplement for cattle and sheep. Drying is both difficult and expensive but it enables the farmer to dispose of this waste product satisfactorily.

With the modern knowledge of nutrition, birds are so well developed that they can come into lay at about 5 months of age, and continue to lay daily for almost a year without moulting, and produce over 300 eggs—a production level which suggests that nutrition and housing are good. Up to ten thousand laying birds can be looked after by one person, provided the greatest use is made of mechanical aids for feeding, watering and also for the removal of eggs and of manure.

The above has been written with chickens in mind but much could apply equally to turkeys, though the dimensions for pens must allow for the bigger birds. Cages are different because the laying turkey hens are not kept in battery cages but in special breeding pens. So far, ducks and geese have not been housed under intensive factory conditions indoors, though some outdoor intensive systems are practised.

Pigs

Some farmers have kept pigs on factory lines in intensive units, sometimes as complete herds with both breeding and fattening units, occasionally in breeding units alone, but more commonly as fattening units alone where weaners are bought in from small farmers who have breeding units.

Undoubtedly, the simplest arrangement is for the fattening unit to be run on intensive lines with up to 20 pigs per pen and 1,000 pigs in a fattening house. One man will look after 2,000–3,000 pigs if the buildings are well arranged with a modern system of feeding and mechanical means for the removal of manure.

Breeding units are not as easy to organise on concrete as fattening units, for, unless special precautions are taken, there is a real risk of iron deficiency appearing, resulting in anaemia in the young suckling pigs. If this should appear, young pigs may die or make slower growth than normal, but the use of the latest iron preparations will usually correct this deficiency. When the whole herd is housed it is essential to have (*a*) pens for the sows for mating, and in which they can remain till they are about to farrow, (*b*) individual pens for boars to be kept near the breeding sows, (*c*) either farrowing pens or maternity pens in which the sows spend about 24 hours before farrowing and a day or two (up to a week) depending on the need for the pens, after farrowing, before going into (*d*) pens where they will remain till their piglets are weaned. There are many schools of thought; one favours traditional weaning at 8 weeks, which on average results in sows having 2 litters in 13 months. The other plan is an attempt to speed up the breeding rate by weaning the piglets at 3–5 weeks. Since sows will usually breed at about 3 days after weaning, it follows that early weaning should result in quicker breeding, and sows may then have 2 litters within 12 months. Not only does this lead to quicker breeding but it results in keeping the sows farrowing at stated months, thus using the farrowing accommodation regularly and avoiding congestion.

This earlier weaning means that greater care must be taken to ensure the suckling pigs are eating creep feeds as soon as possible. It is usual for the piglets to enter the fattening house at weaning and to remain there until they are sold fat, or until the gilts that have been selected for breeding are removed. In a breeding unit with well-arranged pens one man can look after up to 100 sows, assuming the weaners are removed; but if they are retained until sold for bacon, then he can only deal with about 50 sows, 2 boars and their progeny.

Various pens and houses are needed for the complete herd, and it is essential to provide different temperatures for the pigs at various stages of development. The newly-born piglet needs most heat (35°C), and this is most easily provided with infra-red lights close to the floor of the pen, but away from the sow, so that the piglets are enticed away from the sow for safety. Some breeders have found a system of underfloor heating to be satisfactory. Piglets require extra warmth till about 3 weeks of age, but this need falls away from birth gradually so that, except in very severe wintry weather, no artificial heat is provided after the piglets are 3 weeks of age.

The sows, after weaning and up to the time they are about to farrow again, can be housed in cool houses or pens at 17°C. The fatteners need warm pens, but these can be provided, without artificial heat, all the year round with insulated buildings having special ventilation in hot summer weather. The optimum temperatures for fattening pigs are: for porkers under 100 lb liveweight, 20°C; for baconers, 18°C; for heavy hogs, 15°C.

With the intensive pig "factory" and no farm, dung becomes a major problem. In some instances it has been found essential to use no straw and to build a sewage disposal system for the aeration of the dung so that the waste liquid is so purified that it can pass into a ditch and away from the farm. Several disposal systems have proved satisfactory, but often at considerable cost.

Such an intensive system as described above, being quite divorced from a farm, is dependent upon foods being bought ready-mixed for various classes of stock; usually three mixtures are needed, namely one for the sows, another for the creep feed of the suckling pigs, and a third for the fattening pigs.

INTENSIVE UNITS OF LIVESTOCK AS AN INTEGRAL PART OF THE FARM

Between the two world wars, British farming passed through a severe depression. Farmers often pursued a policy of mixed farming, hoping that something would pay a dividend, but with rising labour costs came the need for specialisation. Later the value of land, to buy or rent, rose; the resulting rise in fixed costs focused attention upon intensification of enterprises. This led to increased levels of manuring of arable land, and to a slightly less extent to temporary and to permanent grassland. This increased growth of grass demanded higher stocking rates and new systems of grazing, to ensure that the extra grass produced was fully utilised. Thus a new era of intensive production from livestock has been developed, particularly during the last 20 years.

There is a growing tendency for a farmer to develop 1 or 2 large intensive units of stock on his farm. The advantages are that he is often growing food for his own stock, and this may make it unnecessary to buy in large quantities, sometimes at high prices. In these cases the stock may make good use of grass and grassland products, such as hay and silage, though in certain circumstances it may be more economic to buy hay rather than to grow it.

There may be supplies of straw available to provide bedding for the stock, but again it may prove more advantageous to buy straw than to grow it. Finally, there is land on which to apply the farmyard manure produced by the stock. The integration of stock with an arable farm is unquestionably good husbandry, as it enables farmers to grow a greater variety of crops, thus avoiding certain crop diseases, pests and weeds.

Various livestock enterprises have been developed on general farms over the years. The main ones are:

(*a*) Dairy cows;
(*b*) Calf rearing for the dairy, for beef production and for veal;
(*c*) Fattening cattle including bull beef;
(*d*) Breeding flock of ewes;
(*e*) Breeding herd of pigs;
(*f*) Fattening unit of pigs (with or without breeding);
(*g*) Poultry unit for rearing, broiler production or egg production.

It is possible for one or more of these enterprises to be found on the same farm; this is more likely to be the case either if it is a large farm or a farming organisation consisting of several farms working in co-operation.

Dairy cows

As a general rule any intensive unit for dairy cows must be on a farm, for in addition to the building, in which the cows live and are milked, there must be provision for growing grass, or a substitute such as green cereals for silage making and for green soiling. All concentrates can be purchased so there is no need to have arable land for the production of concentrates. It is essential to have some land, either grass or arable, on which to put the farmyard manure from the cows. Although in the past "town dairies" were operated without available land, present-day economics make it essential to have an area of land associated with a dairy, whether it is intensive or not.

In Great Britain, the average size of the dairy herd has increased over recent years. Herds of 20–40 cows, many of which were run on family farms, have tended to disappear, because of the "tying" nature of the work and the economics of milk production. Herds of 70 cows and upwards have become far more common, and there is now a trend to establish herds of 120–150 cows. These numbers under modern conditions permit better use of full-time dairy labour. Where the layout and planning is right one man can milk up to 120 cows. Furthermore units of such size give the opportunity to plan a relief system, thus allowing workers reasonable time off at weekends, during illness and at holiday periods.

The congregation of a large number of cows at one centre, while being convenient for the actual milking and feeding of concentrates, may be very inconvenient for feeding silage and for grazing. Taking silage first, it is well known that for self-feeding from a clamp, cows need 6 inches width per head, and experience has shown that the maximum number to feed at one face of the silage clamp is about 120. Thus where larger numbers of cows are to be self-fed one must organise several faces of silage to be fed simultaneously. An alternative system for these large units is to deliver tower silage mechanically, either by an auger system or chain and flight, into a trough that is so arranged that cows may feed from both sides at the same time. In this way a large number can be fed in a batch.

INTENSIVE LIVESTOCK SYSTEMS 111

The latest development is to provide completely mixed rations for feeding ad lib, and so not be dependent upon feeding each cow according to her milk yield.

A major problem arises at grazing time. If more than 120 cows pass through one gateway to a paddock, they will do damage by poaching whenever the soil is wet. It is, therefore, essential to divide the large herd into units of no more than 120 cows. Poaching occurs at two places; at the gateways through which the herd must pass, and as soon as they arrive on the grazing area. The poaching is most serious where the soil is heavy, but on light sandy or chalky soils it is very rarely a problem except in very wet weather; fortunately, at such times, the stock are often housed.

With intensive systems of keeping cows some farmers are so afraid of poaching their grass fields that they have turned to zero grazing in the summer time. This means the cows are entirely housed for the whole year and fed freshly cut grass in the summer and silage in the winter. The equipment that is used for cutting and carting the silage may also be used for cutting the grass and feeding it green, providing the grass can be fed in mangers or troughs from side-delivery trucks.

A further problem arises when a larger number of stock are grazing. They will leave much manure on a paddock, and as a result the grass is fouled. Unless there is some irrigation system to wash away the manure, stock may refuse food at the next grazing.

Big herds, when housed for the winter, and to a lesser extent when brought in for milking, produce a slurry problem. It can be a real problem, either when the cows are eating from the face of a clamp silo, even if it has been built on concrete, or when they are kept in cubicles and their droppings collect in the passages that run at the back of the cubicles. In both of these cases the slurry must be removed frequently, often at least every two days, either to a pit or into a slurry cart for spreading.

When one milker has a large number of cows under his care, the disease problem may, by neglect, become acute. The two most common troubles are mastitis and sterility; yet with an observant herdsman mastitis can be kept down, by hygiene, as soon as it appears. A keen herdsman too will observe when cows are on heat, get them served at the appropriate time, and keep records so that he can tell if they are holding to service.

With big units, it is essential to have first-class cowmen, and they must be well paid for their efficiency. These large dairy units are increasing in popularity, and providing the above problems can be met, it is expected that such units will become the recognised practice.

Calf Rearing for the dairy, for beef production and for veal

On many farms all calves are reared in the same general way. They are put into individual pens when they are a few days old where they are given whole milk or milk substitute by various devices: sometimes from a bucket, at others from a mechanical device into which milk, or milk substitute, is put, warmed and circulated, for the calves to suck. In all cases calves must have access to liquid in some form as milk or milk substitute, or dry food and water. In the majority of cases calves will be offered hay after they are 3–4 weeks of age.

The best results are obtained if the calves almost from birth are kept in individual pens with a dry floor which is produced by either a deep bedding of straw, or a slatted floor with good drainage, or a combination of slats and straw. The building should be warm (at least 16°C) and free from draught. Opinions differ whether the building should have neutral light or artificial light, but what is certain is that there must be adequate light to ensure that the stockman can inspect the stock at least twice daily.

Some may be surprised that it is recommended that calves are to be kept in individual pens. The reason for this is that there is considerable risk that calves may scour from infections, colds, and incorrect feeding, and experience has shown that if one calf scours and it has access to others, then scouring may spread like an epidemic to adjacent calves. Thus, if they are kept in individual pens, fitted with solid divisions, the risk of infections spreading is very much reduced although, of course, air may be contaminated in the same building. The calves entering this house may be home-bred, in which case the risk of disease is very much less than if they have been brought in. Purchased calves provide the greatest risk if they have been bought from a market where they may have picked up infection or colds, in transit and from contact with other calves, while there is less risk if they have come direct from other farms. In practice both systems are to be found. As a general rule, bought-in calves should be given, on arrival, warm water containing glucose and it would normally take 24–48 hours to turn them from the glucose ration to a proper milk, or milk substitute, feed. With this sweetened water there is much less risk of scouring from indigestion. There is less risk of the calves bringing in infection if bought when 14 days of age than at 2 days, since the former will usually have had colostrum from their own dams before they have left the farm on which they are born, and this is essential to give them a healthy start.

Some calves bought in for beef production from dairy herds may be reared in a different way: in groups of 20 on a special supply of milk, or substitute, from a "Mechanical Udder" with teats. Experience has shown that this is a very satisfactory system of rearing, since the calves can drink, at will, electrically warmed milk (or milk substitute) almost from birth until about 3 months of age. In this way one man can look after various batches of calves, numbering 200–500 in all, depending on arrangements.

Calf mortality is a very important problem. Even in well-run herds one hears of losses of the order of 20 per cent. This arises from a combination of cold, draughty buildings and calves chilled (and possibly starved) in transit, resulting in indigestion; this affords the ideal conditions for *B. coli* and Salmonella to thrive, and the inevitable scouring appears. The resulting bacteria contaminate the building and an epidemic follows. The only certain way of stamping out such an infection is to remove all cattle, remove litter, disinfect the building thoroughly, and leave it empty with doors and windows open for 2–3 weeks. It is quite apparent that such outbreaks must be treated very seriously, otherwise the infection will persist in the buildings and epidemics will appear periodically.

When the calves cease to have liquid milk, or liquid milk substitute, either they may be put into yards (with 20–50 calves in a yard) or the partitions may be removed to allow the calves to run together. At this early age it makes for easier management if all in one pen are of the same sex and approximately the same size.

Ideas have changed regarding feeding fat to calves. Formerly they were reared on skimmed milk, with carbohydrates added only to replace the milk fat that had been removed. More recently the fat content of milk substitutes has been raised so that it has now reached 24 per cent, the outcome of which has been quicker liveweight gains of the calves and less nutritional scour. These emulsified fats have been sprayed or blown into the calf mixtures by the cake manufacturers. Since the smaller the fat globules are the greater the speed at which they can be digested, it is apparent that the addition of the fat is a factory process and is unsuitable for home-mixing.

Veal calves are usually kept in individual pens from birth, in a temperature of 23°C, until they are sold fat when about 3 months of age. During this time their diet is milk, or milk substitutes, and often nothing else at all since other foods result in the flesh being less white at slaughter and so of lower value.

Calves for the dairy herd, after they have been in individual pens, are either moved into other pens or into yards in groups of about 20–40, at the age of about 6 weeks. In such pens they receive concentrates, water and hay ad lib, and if it is available, silage as well. These calves in boxes or yards do not require as much warmth as the young calves, and will do with under 15°C. They may remain in such pens until they are turned out for grazing in spring. One man is capable of feeding and managing several hundred calves if buildings have been carefully planned. Small pens demand more labour for feeding and littering than larger pens, and hence the variation in number of stock per man. Dairy heifers, after they are out on grass, may never be housed again until they are brought in for calving.

Turning to beef calves, they may be reared as above but with the main difference in procedure, compared with the dairy calves, that they may never go out to grass and they are fed ad lib all their lives until they are sold fat at 12–15 months of age. They are kept in groups of about 20 until they are sold. Other beef cattle, if they are not fattened continuously from birth, may go out to grass in the appropriate spring and then be housed when the grass deteriorates; depending on district and season, this will be in August or September. They are then fattened on liberal rations of silage (or roots), hay and concentrates. They are kept in pens of about 20 per pen, and one man can look after up to 1,000 if the buildings are well arranged and the feeding is well organised.

Fattening cattle, including bull beef

It is often convenient to house fattening cattle in groups of 20, matched for size and sex. Larger numbers per pen usually result in bullying at feeding time, and the smaller nervous animals may be unable to obtain their allowance of food. Further, if larger numbers are kept together the fattening is not uniform throughout the whole group, and then stock may be sold fat over a long period; this means that at times the yard is not kept full. If cattle are introduced from another yard it is unsettling for all, but if few beasts remain in a yard that is not completely used, this is very expensive from a housing point of view.

If there is a convenient central feeding passage with pens for 20 stock off that passage, then one man is able to feed up to 1,000 cattle without any special problems providing feeding can be done mechanically either by tractor and trailer or by some auger system. It is obvious that any arrangement of this kind must afford provision for taking stock out of pens (when they are off feed or ill), and for weighing them to ensure that they are reaching the required weight for sale. As a general rule there is no slurry problem because straw may be used freely for litter. This can accumulate under the stock for 6–12 months if the yards have been constructed with sufficient depth and fitted with rising mangers. In such cases the farmyard manure may be removed from the yards after the stock have been sold, and then the yards may be prepared for the next batch for fattening. In some instances the same yard is used continuously for the same batch of cattle until they are sold fat at 12–15 months.

In the early 1960's, a new system of fattening beef cattle was developed which consisted of putting calves, when about 3 months of age, on to an ad lib ration consisting of nothing but rolled barley until sold fat at 10–12 months of age. In this way some batches of calves made excellent liveweight gains; this was particularly the case with some families of British Friesian steers, while others were disappointing. Poor results may have been due to breeding, with this feeding system, resulted in digestive disorders; the worst of these resulted in death from bloat, for the crushed barley ration suited some animals but not others. This original ration has now been changed from entirely crushed barley to a basic ration of about 2 lb of hay per day and a cereal ration given ad lib consisting of crushed or rolled barley, oats or wheat depending on prices and availability. This small ration of hay has prevented bloat and assisted the digestion; this has now become a safe, recognised feeding system. In addition a protein supplement may be used, which also includes minerals and vitamins if required.

Not all farmers wished to keep their cattle entirely housed all their lives, so they developed the following slightly different system. The autumn-born calves are fed liberally from birth till the grass has grown sufficiently for them to be turned out to graze in the spring, often to new ley (this reduces the parasite risk). They are given a little hay to prevent scouring while getting used to the young grass. When acclimatised, their sole diet is grass and they remain grazing until the grass shows signs of deterioration in September. Then they are brought in and fed on a crushed cereal ration with a basic allowance of hay, and, if available, silage. These steers are then about 12 months of age, and weigh about 6 cwt when housed for fattening indoors for a further 3–6 months. They are sold fat, weighing 9–10 cwt, between Christmas and Easter, when beef prices are at their peak. There is a good demand from consumers for this semi-intensive beef, and it suits farmers because some grass is grazed with this system.

Bull beef

The most recent development in Great Britain is to copy the Continental systems of fattening bull calves, that are never castrated. Providing they are sold before they are 12 months of age (i.e. before sexually mature), they produce good quality, and dark-coloured, lean meat that has been quite popular. Such calves are housed all their lives because there are risks in having such cattle grazing out-of-doors. Bulls can be kept in this way only if special permission is obtained from the Ministry of Agriculture. A bull that is not fat before 12 months of age may become a nuisance if he remains on the farm.

Breeding Flock of Ewes

As a general rule, most flocks of breeding ewes remain out at grass all the year round; but during the last few years, there has been a growing tendency to house the ewe flock in cold areas, or where the soil is heavy and there is a risk of poaching. This poaching is not only bad for the sheep, but is also bad for the grass, because it leads to very late growth of grass in the spring and it may also lead to the elimination of good grasses and clovers. On such farms the policy has been to house the ewes from about Christmas to Easter, a period of 3–4 months. This means they may be housed during the latter stages of pregnancy and for a short time after lambing.

Usually no special permanent buildings have been erected for these sheep, but frequently the policy has been to use existing yards or to build cheap temporary buildings made of polythene. Sheep do well in straw yards, and if straw is in short supply then slatted floors are a great advantage. The sheep lie dry on such floors and their fleece will be kept free from straw.

It is possible for one man to look after 1,000 sheep under cover in this way providing they are in groups of about 100 to 250. Grouping will enable the shepherd to divide the flock into ewes to lamb, lambed ewes, ewes with singles, ewes with doubles and ewes with more than doubles.

Some farmers have a very different policy and keep their ewes in groups of 50 per pen, so that ewes lamb over a period of 3 weeks. They find that with a very few ewes lambing in the pen at any one time, there is less chance, or opportunity, of mis-mothering, and consequently the ewes take better to their own lambs.

The food of such housed ewes is normally hay, concentrates and possibly roots or silage. The lambs will eat much the same foods as their mothers, with the exception that slightly better quality concentrated foods may be provided in creeps for the lambs. All housed sheep must have access to water. This intensive housing has produced good results with lambs, but some find when shearing that housed ewes have slightly thinner coats than similar ewes kept out of doors, and have clipped up to 2 lb less wool per head than outdoor sheep. In general 6–15 sq ft of floor space must be allowed for each ewe, depending upon breed, while half this is allowed for each lamb. Some farmers allow $12\frac{1}{2}$ sq ft per 160 lb ewe.

Two other problems have arisen. One is foot-rot in the sheep. To avoid this it is essential to keep the floor as dry as possible and it may be essential to run sheep through a foot bath periodically. The other problem is that of parasites. This, in the past, has often prevented farmers from housing sheep, but providing the ewes are dipped before they are mated, the risk of parasites such as ticks is very much reduced.

With intensive sheep production, one needs a big lambing percentage which means many twins and some triplets with the ewe flock run on traditional lines. Ewes cannot feed more than twins well, and so the third lamb from triplets must be either fostered or hand-reared. The latter is time-consuming and is most unpopular with shepherds; it usually results in the third lamb dying or growing very slowly. The housing of the flock makes artificial rearing of the third lamb from triplets relatively easy and profitable. The system is:

1–2 days after birth	—Colostrum from ewe
3–10 ,, ,, ,,	—Feed from a teat with 2 pints of milk substitute* in 4 feeds. Also taught feeding from shallow bowl.
10–30 ,, ,, ,,	—Change gradually to weaner pellets* and water, also with grazing and warm housing.
Over 30 ,, ,, ,,	—Wean from milk substitutes to concentrates and out to grass.

* Commercial mixtures that can be recommended are available.

These lambs also may be fattened indoors with more rapid gains, on a diet of concentrated foods only.

One variation is to lamb ewes indoors, to wean the lambs when they are 2 months of age, to keep the lambs indoors, and to fatten them as rapidly as possible on concentrates, hay and roots. With lambing in January, the lambs are sold fat when prices are high in March and April.

A further variation occurs with lambs born out-of-doors. They may be weaned at 2 months of age and taken indoors for fattening on a diet of concentrates fed ad lib until they are slaughtered when weighing 80 lb liveweight and not more than 4 months of age.

Finally, shepherds like this system of housing ewes at lambing time, for it makes their work much easier, especially if lambing early in the spring during a period of bad weather.

Some farmers remember the days when ewe flocks were kept on arable land and were lambed in permanent lambing pens; disease was common and losses of both ewes and lambs were high, arising from the accumulation of infection in a relatively small area. So far this revival of indoor lambing, provided the fullest use is made of modern hygiene, has not created undue disease problems. In fact excellent results have been achieved and this may be because annually the yards are cleared out thoroughly.

Breeding herd of pigs

On a number of farms a breeding herd of pigs may be kept as a one-man unit. If a man does nothing but look after the breeding pigs until the little pigs are weaned, he may look after about 100 housed sows with 3 or 4 boars, but for this to be achieved the buildings must be well arranged. The number may be doubled with some outdoor systems; these may be practised in the summer everywhere, but only in the warmer areas in the winter time. The housing has been described earlier in this chapter (p.107).

The advantage of having the pig breeding unit on a farm is obviously very great because there may be a supply of straw for litter, and also because of the ease at most seasons of the year when waste effluent can be spread on the land. Farmers may have a herd of 50 sows and fatten all stock until they reach bacon weight; this makes a very convenient one-man unit. For various reasons the farmer may have breeding at one centre on the farm and fattening at another. On the other hand on many grassland farms some of the pigs will be out-of-doors, namely in-pig sows and sows just after farrowing until weaning. In warm districts this policy is adopted throughout the whole year, but in colder districts in the north of England, or in Scotland, all suckling sows must be housed between Christmas and Easter.

Since as a general rule sows may have two litters in 12 months, the policy is often to farrow at four main periods of the year: half of the sows will farrow in January and July, the other half farrowing in October and March. In the past, experience has shown that suckling pigs out on grass, providing they are on new grass that has not been contaminated by previous pigs, will make better liveweight gains than housed pigs. This was so before iron preparations were available to prevent anaemia in housed suckling pigs; but now they are available, farmers often obtain as good results with stock entirely housed as with suckling pigs at grass. Of course, if one has a large number of pigs on a small acreage of grass or ley, and it is then ploughed up, heavy pig manuring will produce big crops of roots or potatoes and this may prove a sound farming policy. Pigs fit in well into an alternate husbandry system, especially on the lighter soils.

Fattening unit of pigs

Some farmers with large arable farms have fattened pigs from weaning until they reach bacon weight of 200 lb, or heavy hog weight of 260 lb, by keeping them in large units in straw yards or pens. This has proved to be a very satisfactory way, at a low cost, of utilising the covered yards that were originally built for cattle. It also provides a very good way of utilising straw and of making farmyard manure for the farm.

Some arable farmers who do not wish to fatten the pigs themselves, may let their buildings to others on the understanding that if the farmer gives the straw, he will receive the pig manure. This arrangement is sound husbandry because it means that farmyard manure will be made in large quantities for application to the arable land. When such pigs are brought in direct from other farms, to ensure that pigs of the right type are produced the farmer purchasing stores may supply the boar to the farmers breeding the stores. On arrival purchases are usually dosed for worms and treated for external parasites, inoculated against erysipelas, and then fattened on the conventional rations of concentrates and water.

In some districts, however, where there is a good supply of skimmed milk or whey, pigs may be given such liquid food mixed with meal. This modern method of liquid feeding (in which liquid is pumped to the pens) is gaining in popularity, since it demands less labour and enables a very large number of pigs to be fed by one man in the minimum of time (it is even being done by computer). With such modern arrangements, one man can feed several thousand pigs. The pigs must be weighed expeditiously at weekly intervals as they approach the desired weight for bacon or for heavy hogs. To do this the pigs must be controlled for weighing and be returned subsequently to their respective yards or pens. They must be marked and easily identified, and selected for sale at the right weight or time.

Experience has shown that when a large number of pigs are kept some must be sold each week, otherwise they will not be able to obtain maximum prices. Bacon pigs should have restricted feeding after they have reached 140 lb in weight, but with heavy hogs ad lib feeding will usually give the best returns.

Poultry units for rearing or broiler production or egg production

As a general rule, poultry units have not been found on general farms but they have rather been put into special units; in a few cases, however, they have been put on general farms. They are usually run by a specialist, and the only services that are given by the farmer are to provide litter and remove manure. These are the only effects the specialist enterprise has upon the farm. The only exception to this is that a deep litter unit may utilise the existing yards—adequately fenced, of course, for the laying birds—but this system is not very common.

KEEPING LIVESTOCK IN INTENSIVE UNITS

This presents a number of problems.

Firstly, where there are large numbers of stock there is a great risk of diseases accumulating and of pests appearing. These risks are reduced if the stock are all home-bred. Most troubles appear if stock are purchased, especially if from various sources including the local stock market. Thus if one has to buy a large number of stock, one must take precautions when they arrive on the farm. They should be put into isolation pens and dosed against internal and external parasites, with pigs inoculated against erysipelas, and with special precautions taken against scouring in young cattle and sheep.

The second point is that when one is rearing a large number of young animals the risks are very great of troubles (diseases and pests) appearing. When stock numbers are large, the troubles may appear as epidemics. As a wise precaution, after each batch of young stock has been reared in a building, that building should be closed; all dung, etc. should be removed, and the building then disinfected and left empty, with liberal ventilation, for several weeks. This means rests after each 3 months when calves have been reared, and after each 4 months when pigs are fattened, but only after each 10 weeks with fattening broilers. Thus instead of rearing 5 batches of broilers in the year, one can only run 4, when one allows for 2 weeks after each batch. The disinfection must be thoroughly done with power units to wash into cracks and crevices where pests and parasites may otherwise accumulate.

The third problem arising with large units of stock is that labour must be efficient. Whenever a man or a woman is required to look after a large number of stock, the stockmanship must be good and great attention must be paid to detail. All stock must be seen daily, preferably twice daily, and the person in charge must be conscientious and carry out his, or her, duties completely. With large numbers it is essential that individual attention should be given, and that very great care is taken to ensure that troubles are discovered before they can become epidemics. The person in charge must be exceedingly observant and pay particular attention to hygiene where young are born and where young are reared, though perhaps to a less extent where older stock are kept; in all cases standards of hygiene must be high.

A good stockman is ever-watchful for the first signs of sickness, which are usually an animal or bird off its feed followed by a lack of condition and disinclination to mix with other stock. Of course, if breeding is involved, the stockman must be constantly watchful so that stock are bred at the proper time. A good stockman can make all the difference between success or failure when a large number of stock is being kept. With dairy cows sterility may not be noticed, and it is only when the cows have failed to calve regularly and consequently a bad calving index has appeared that the trouble is observed.

Some farmers, to encourage their stockmen, give special incentives in addition to a wage (which may be dependent on ordinary and overtime worked), such as bonuses for calves, lambs or pigs reared. Other farmers prefer to give a high standard wage with a bonus on the overall profitability of the unit.

While mentioning labour, other problems must be raised. In manufacturing industry there is a growing tendency for employees to work 5 days a week, and the same policy has appeared on arable farms. It is now accepted that stockmen must have some weekends off duty. A new policy may be adopted by farmers, namely for their stockmen to work for four or five days, and then for them to have a free day, not always at the weekend but on a rota basis. This has worked well in some industries, and it may have a place on the farm with livestock where several stockmen are employed; but on a farm consisting only of a one-man unit the relief depends on the farmer himself taking his turn of stock duty, unless a relief worker comes in from another farm or a relief service.

Further problems with pig, calf and poultry units are that with housing there must be complete control of temperature, ventilation and humidity for the younger birds and animals, from birth or hatching until several weeks of age. Young stock must have priority of accommodation.

All stock must have a dry bed; this is possible only if there is provision for drainage of urine and cover to keep off rain. All housed stock must be fed quickly and easily either by some mechanical means or by the stockman himself. This is most easily done if foods can be given without the stockman entering the yard, pen or box. While feeding is taking place, the stockman should be able to observe the stock, to isolate sick ones, and if necessary to dose them. Buildings should be arranged so that manure can be removed easily and sometimes mechanically, either throughout the whole year or periodically, depending on the policy that is being adopted.

From time to time it is necessary to move the stock from one place to another to regrade them and to weigh them. Complete control is necessary at all times, and especially when stock have to be isolated for breeding, treatment or dosing. In the latter case it may be an advantage to have special pens, with special equipment to restrain stock.

Finally, it must be easy to isolate stock, in order to cleanse buildings and remove manure after a batch of animals has been through the buildings, or after troubles have occurred. Thus the more that walls and floors can be constructed of concrete, the easier disinfection will be. Wood, on the other hand, while being a cheap material for construction, is difficult to disinfect.

CONCLUSIONS

Before a farmer can decide upon the enterprise to follow on his farm, he must investigate supplies of food and stock and be sure that he can attract adequate labour to carry out his enterprise. Having produced the stock, he must be sure of the sale of produce that will command a good price. The best prices are obtained if there are regular sales rather than spasmodic ones. This means that if a farmer can be in a large way of business regarding one particular enterprise, he has a greater advantage over his fellows, who have fewer stock and therefore cannot have constant sales. It will be seen, therefore, that the larger farmer, or farm with large enterprises, definitely has a great advantage unless farmers form themselves into groups and co-operate to sell stock and, of course, for the purchase of foods and manures.

Chapter 8

FRUIT, FLOWERS AND VEGETABLES

In this chapter, by way of contrast with the previous chapter, we pass from intensive livestock production to intensive crop production. Fruit, flowers and vegetables are produced in many parts of Great Britain. In selected areas, where soil and climate are suitable, fruit and flowers are grown in the open; but in many places, where the climate and soil are not entirely suitable, glasshouses are erected, and this has been especially so near densely populated areas. Often fruit, flowers and vegetables are grown around the coast, where soil is suitable and the climate is temperate. In these cases the produce will be ready for sale earlier than that grown from similar soil further inland. Examples are the Channel Islands and the areas around Penzance, Bournemouth, Romney Marsh, the Isles of Sheppey and Thanet, the mouths of the Rivers Orwell and Deben, Waveney and Yare, The Wash, the Humber, the Tweed, the Firth of Forth and the Firth of Tay, the coastal areas of Ayrshire, Lancashire, North Wales (especially around Anglesey), Pembrokeshire, and, in Northern Ireland, County Down and Armagh.

In several of the river valleys nursery crops, flowers and vegetables are grown, the more important being those of the Thames, the Colne, the Stour, the Waveney, the Ouse first in Bedfordshire and later in the Fens, the Avon and Severn valleys (Evesham and Worcester), and finally the Parrott in Somerset.

SOIL AND CLIMATE

For all these crops, soils must be free-draining. Artificial drainage may be necessary for vegetables and flowers, but this may not be so effective for fruit trees where there is the danger of roots blocking drainage pipes. The soil, however, especially for trees and bushes, should have sufficient body in it, to hold moisture to withstand periods of drought. For flowers and vegetables, irrigation may be carried out, but for this the soil should not readily cap on the surface, as this would make irrigation difficult, or impossible. Then, for trees and fruits the soil should contain sufficient lime or chalk, which means a pH of not less than 6·0 but not over 6·5 for apples or pears. For the various flowers and vegetables, deficiencies may be remedied to suit the needs of the crops. In all cases a soil that is naturally high in organic matter is desirable, for if it is not there naturally, it must be added artificially.

Other main nutrients can be added as needed, but the more that are there naturally, the cheaper it will be to grow the crops. It follows that where crops are grown in rotation, the nutrients will be given most freely to those crops that need them most and lighter dressings will be given to the other crops. When minor elements are involved they will be given in small quantities, more frequently as sprays and not as manures.

Glasshouse crops are less affected by soil and climate than outdoor crops. The produce is so valuable that the soil can be transported for short distances to such sites. Formerly it was essential to change the soil frequently because of the accumulation of pests and diseases, but now this is no longer necessary because the soil can be sterilised periodically and so be retained for an indefinite period. With artificial heating and humidity control, it is possible to produce the right conditions for the production of a wide range of flowers and salad crops throughout the year, but if the house is in a cold locality then the cost of heating will be high. Now carbon dioxide is forced into some glasshouses to increase growth rate.

Horticultural crops may run the risk of drought affecting yield and quality. Rain is the cheapest source of water, but it is not always reliable; so the nursery must have a reliable supply of irrigation water from rivers, wells, reservoirs, or mains supplies. It should be noted that licences are required in Great Britain to use water for irrigation, and may be refused to new users; in general, charges per 1,000 gal are made for all water used for irrigation purposes.

Turning to outdoor fruit, climate is very important. The greatest risk is late frosts in spring when the fruits are in blossom. Any area selected for fruit growing must be relatively free from such risks. Consequently fruit growing is found in selected areas, particularly on the south slopes of hills affording a sun trap, and avoiding hollows where frosts are more likely to occur. To a limited extent, frost-free areas are chosen for early vegetable crops; but since these crops are often grown on a field-scale, areas are sometimes used (a) where there are occasional risks of spring frosts, which may delay drilling or affect germination, and (b) where early autumn frosts may damage the crop before it is cleared: this is particularly risky with a crop like celery, which is tender and susceptible to frost damage in late autumn.

PRINCIPLES OF PRODUCTION

The general principles of production can be discussed under the following headings: (a) labour, (b) cultivations, (c) manuring, (d) chemical aids, and (e) marketing.

Labour

Fruit, flower and vegetable enterprises formerly demanded much labour, and so were developed in areas where labour was freely available. In recent years, however, owing to high labour costs and shortages, labour-saving methods have appeared. Much of the hand-hoeing and weeding has been superseded by herbicides. Although some of the harvesting has been mechanised, many crops still make heavy demands on hand labour. For example, the preparation of various plants for sale and the bunching and boxing of flowers demand very careful handling. The picking of such fruits as raspberries, blackcurrants, apples and pears is now partly mechanised, but the picking of strawberries and of tree fruits such as cherries and plums all demand hand work. Thus the siting of these enterprises is important, because if labour is not available on the farm itself, or in a neighbouring village, special arrangements may be essential to transport gangs daily for picking and marketing fruit, and for harvesting, grading and marketing some vegetables.

To meet modern demands, grading and packing (and sometimes washing) of vegetables and fruit is essential if maximum returns are expected. Some of this work is mechanised, but even then much hand labour is still required. Preparation for market is sometimes done completely on the farm, but it may also be at a co-operative collecting station or "factory" to which the produce is taken for grading and packing after some of the preliminary work may have been done on the farm. This latter procedure retains waste material on the farm, some of which may be fed to stock. In order to achieve orderly harvesting and to make the fullest use of machinery, some farmers are organising themselves into co-operative groups to market the produce after it has been graded and packed.

The production of fruit, flowers and vegetables in the United Kingdom was originally centred around the large consuming centres, and growers themselves hawked their produce direct to consumers in the towns and villages; alternatively they sold the produce from stalls in the markets in consuming centres. While some of these customs still persist, in most cases there has been a major change. The small producer-retailer has very largely disappeared, and in his stead one finds the wholesaler and retailer of fruit, flowers and vegetables. Thus the growing of such produce is not of necessity near to the consuming centres, but developed in areas where soil and climate have been ideal for production. Modern road transport has also had a significant effect. In all cases, however, labour must be available to grow, to prepare the produce for market and to dispatch the produce.

The demands for labour vary tremendously from farm crops to the fruits, flowers and vegetables. This may be demonstrated by the few examples given below.

Standard labour requirements in man-days per year per acre of crop are

Wheat	1·5
Potatoes	12·6
Peas—pulling	35·0*
Orchard (culinary)	40·0*
Small fruit	50·0*

Another way of showing the big differences in labour demands may be seen from the fact that, whereas on large arable farms one man can work over 200 acres per annum, on a mixed holding with grass, farm crops and a good proportion of horticultural crops it may fall to 15 acres per man, and with only glasshouses to 1 acre or slightly more per man (or woman). In all cases, during recent years, economies have been made in labour demand, arising from the high labour costs which accompany a shortage of labour.

In fruit, flower and vegetable growing, seasonal casual labour is employed (this has been allowed for in the figures) and may reach 20–25 per cent of the total labour force over a year. The actual months in which this casual labour force is needed will depend on the crop, but it will be needed most, in general, from June to October, inclusive.

* If labour costs are high, gross margins per acre may be high also, e.g. vining peas £55, runner beans on sticks £160, and summer cauliflower £180; while Cox's dessert apples may reach £250, and strawberries and plums may be even higher.

Cultivations

Depth of cultivation is most important. For the root crops, usually the land is ploughed to ensure good conditions for root development and periodically it may be subsoiled to assist drainage. It is also important to plough in crop remains and farmyard manure as a means of adding humus to the soil. Although the plough is a first-class implement for deep cultivation, its place may be taken, in some circumstances, for shallow and rapid working by a rotavator. This implement will work the soil thoroughly to a depth of up to 8 inches, though 2–3 inches is often adequate. This enables the grower to make a seed bed quickly for drilling or transplanting another crop on land, after one has been removed. In many crops, weed control formerly achieved by hoeing, is being replaced by the use of herbicides which are often more effective; and furthermore with herbicides the soil does not dry out as readily as with cultivations.

In recent times there has been a decided change from the transplanting, on a field scale, of brassica crops such as broccoli, cabbages and brussels sprouts, to direct drilling with a precision drill followed by singling and, if necessary, hoeing. Direct drilling, on the whole, may be slightly cheaper from the point of view of labour, but it uses more seed per acre cropped: with the risk of unsuitable weather for transplanting, a more certain stand of crop is obtained by drilling.

Mechanisation has influenced the row widths of vegetables; carrots and brussels sprouts may be given as examples. Formerly carrots were grown in rows 12 inches apart; but to suit tractor wheels, the distance has been increased to 20 inches, with the seed drilled in a band about $2\frac{1}{2}$ inches wide, or in 2 or 3 narrow rows. A method which has been adopted in recent years is to scatter the seed on minibeds 9 inches in width. Not only does this method give a good plant population, but it also produces even-sized, well-shaped carrots which are now in demand for the canning trade. In these ways, the plant population has been not only maintained but increased. Another variation has been to have varying row widths, wide only where the tractor wheels are to pass. Another variation, arising from the use of herbicides without hoeing, is to broadcast the seed; but while this increases the total yield, it may lead to complications at harvesting if mechanised systems are used. The volume of the crop to be lifted must not exceed the capacity of the harvester.

Brussels sprouts were formerly transplanted on the square at 36

FRUIT, FLOWERS AND VEGETABLES 129

inches × 36 inches, but for mechanical picking the spacing has been reduced first to 27 inches × 27 inches, and then to 21 inches × 21 inches till with some varieties 18 inches × 18 inches has been reached. This has produced small sprouts suitable for freezing. To meet the needs of modern harvesting equipment for this crop, the use of well-graded and carefully selected seed is essential to give uniformity of germination and subsequent growth. The complete crop must be ready for harvesting at one time, since it is not possible to traverse the field several times to harvest parts of the crop as it matures.

Manuring

Several general principles of manuring must be followed when growing fruit, flowers and vegetables.

(1) All land used for growing these crops, with only a few exceptions, must contain an adequate lime content. The most acid-tolerant crops are potatoes, cucumbers, tomatoes, marrows, rhubarb, strawberries, raspberries and pears (they need a minimum of 5·0 pH), while spinach, cauliflowers, onions, leeks, parsnips and lettuce are the most demanding with a pH of 6·0 to 6·5. If a chemical test shows that lime or chalk must be added, it is most important to apply to the surface of the field only the recommended quantity, for an excess may lead to a trace element problem such as manganese deficiency. Whilst lime may have to be applied for a range of crops, it must be remembered that others, like Rhododendrons and Azaleas, thrive better on acid soils than on alkaline soils.

It is well known that cherries and other "stone" fruits demand a soil with a high calcium content, so unless there is a reasonable amount of lime naturally present they are better grown elsewhere.

(2) As a general rule, in order to grow good crops of fruits, flowers and vegetables there should be ample reserves of humus in the soil. This is often supplied in gardens as farmyard manure. Humus helps the texture, making the working of the soil easier, provides a steady and constant flow of nutrients and assists in holding water in the soil. For many brassica crops 20–30 tons FYM per acre may be required (if it is available), while for potatoes a lesser quantity may suffice. Farmyard manure may be supplied direct to the brassica crops, but for some of the roots (carrots and parsnips in particular, because the manure leads to fangy roots), for legumes and for flowers, it is often better to apply the manure in the rotation at least one year before these crops are to be grown.

(3) The absence of nitrogen can be clearly seen by the presence of thin shoots, with smaller-than-average leaves of a pale green colour. Nitrogenous top-dressings should be given to brassica crops and to a lesser extent to root crops. Often 60–180 units per acre of N as sulphate of ammonia (or its equivalent) are given, but it is very difficult to give precise figures since so much depends upon the soil, the past cropping and manurial policy and the other nutrients that are being applied to the crops in question. Care must be exercised to avoid giving excessive dressings, since they lead to (*a*) too vigorous growth of leaves, (*b*) absence, or a shortage, of seed development (leguminous crops), (*c*) less flowers and a shortage of fruit and (*d*) products with a short keeping, or shelf, life.

(4) The use of phosphatic manures (at 40–80 units per acre P_2O_5) leads to the promotion of root development and is required by all vegetables, including legumes. Adequate supplies of phosphates are essential for all the fruits, whether tree or bush, but in the case of the latter longer-growing crops, manures are applied every two or three years and not annually as for vegetables.

(5) Potash manures applied at 40–180 units per acre K_2O assist in producing quality and colour of produce, points which are most important with eating apples and flowers. It is claimed that the liberal use of potash provides crops with resistance to disease. Thus potash manures are needed by all crops: flowers, fruits and vegetables. In the case of flowers and vegetables, annual dressings are essential; though with tree and bush fruits, applications in alternate years may suffice. On clay soils there is usually sufficient potash present for some of the fruit trees to be grown without the application of potash manures.

(6) There are a few instances of other deficiencies which need to be corrected. Iron shortage, like nitrogenous deficiency, leads to yellow leaves (especially the younger ones) on soils with a high pH; this may appear in raspberries, strawberries, and most of the tree fruits. On the lighter soils, copper deficiencies, resulting in die-back of shoots, are found, most markedly in the cases of onions and pears. Manganese, shown by premature yellowing and browning of leaves, may be deficient on the fen and peaty soils, and especially when chalk has been added in liberal quantities to give a pH of over 6·5. Boron deficiency may cause a disease in some crops known as brown heart, and must be corrected by applying borax. Magnesium deficiency can affect a range of crops causing leaf chlorosis and must be corrected by applying a fertiliser containing a magnesium salt. Molybdenum deficiency may prevent curd forming in cauliflower. Sometimes fruit trees may seem to be quite

FRUIT, FLOWERS AND VEGETABLES   131

satisfactory until their roots reach a certain depth, and then quite suddenly signs of deficiency may appear because the roots have reached a deficient area in the subsoil or rock.

This whole problem is complex, for apples grown on certain rootstocks may show deficiencies which do not appear when the same varieties of apples are grown on other rootstocks in the same soil.

Trace elements are usually most easily corrected by foliar spraying, since this corrects the deficiency rapidly for that crop alone and uses a very small quantity of minerals that may be expensive to purchase.

(7) As a general rule, complete artificial manures are applied to vegetables and fruits but more discrimination is used when growing flowers, special care being taken to keep the supply of nitrogenous manures relatively low for them.

CHEMICAL AIDS

Spraying

During the last 20 years, far-reaching changes in the use of chemicals for the production of healthy fruits, flowers and vegetables have completely revolutionised the growing of these crops. In the use of sprays, four major aspects can be identified in addition to correcting deficiencies, namely: pests, diseases, weeds and arrested growth.

(1) It is recognised that many pests and diseases can be controlled by spraying with the appropriate chemical at the correct time. The development of the crop, pest and disease must be just right if the spraying is to be successful. Some examples may be given from the vegetables, fruits and flowers that are under discussion here.

(a) *Controlling pests by spraying and/or seed dressings*

    (i) *Aphids* may be sprayed when they are seen to be active on the crop.

    (ii) *Wireworms*. Where tests show this pest is present, either powders or liquids are put into, or on, the soil before a crop is drilled or planted.

    (iii) *Leatherjackets*. With this pest the growing crop is sprayed.

    (iv) *Carrot fly*. Seed is dressed with an appropriate insecticide and then the land is usually sprayed after drilling but before the crop emerges.

(v) *Winter moth* is controlled by spraying trees in winter when the buds are dormant.

The actual sprays or powders used are so frequently changed that one advises readers to consult the Ministry of Agriculture's Approved List, which is revised annually.

(*b*) *Controlling diseases by spraying and/or seed dressings*

Seed dressings are used on the various vegetables and farm crops.

Potato blight is fairly well controlled by periodic spraying of the growing crop, and later by killing the haulms to prevent disease reaching the tubers.

Botrytis is a disease, on crops grown in glasshouses, that is controlled by spraying.

Again the reader is advised to consult the Ministry's approved list of products.

(2) Chemicals have been used to control weeds in the growing of fruit, vegetables and flowers. Pre-emergent sprays have been most used successfully, and have reduced the need for hoeing; this has prevented loss of moisture in dry times. Further, the use of post-emergence sprays has played a big part in growing vegetables; thus the appropriate use of chemicals has "invaded" the growing of fruit, flowers and vegetables.

There is one serious problem that has arisen regarding the use of chemical sprays, namely that they have sometimes had very serious effects on bees and the bee population. Bees play a very important part in cross-pollination of fruits and broad beans, so much so that at times absence of bees has reduced production and some farmers have been forced to hire the use of hives from apiaries. To avoid damage to bee stocks from the use of poisonous chemicals appropriate care in their use must be exercised.

The choice of chemicals for weed control depends on the crop and also on the time at which the particular weed or weeds are most vulnerable to attack by chemicals. Briefly the position may be summarised:

Potatoes are either pre-emergence sprayed or sprayed at the end of the growing season to kill tall vigorously-growing weeds.

Carrots and *Onions* may be pre- or post-emergence **sprayed with** appropriate sprays.

For the latest information see Ministry's approved list.

(3) For many years farmers have sprayed potato tops at the end of

the summer, to arrest the crop's growth to prevent the tubers becoming too large for seed, to ripen off the crop prior to lifting at harvest, and, in some instances, to prevent potato blight disease from spreading from the leaves to the tubers.

(4) There is yet another reason for spraying that is still in the experimental stage, namely the use of a growth-regulating chemical to arrest the growth of brussels sprouts' stems in order to force the development of small buttons (Quality Size) close together on a short stem. This should increase the production of high quality sprouts, and makes mechanical picking easier as the maximum number of sprouts are ready for picking at the same time. Some growers achieve the same result by removing from each plant the central growing tip either late in August, or early in September, according to the advice of the factory representative, if growing under contract for the quick-freeze trade. A similar effect may be obtained by growing the new hybrid varieties that have recently appeared, e.g, Atlas, Mercator, Erika and Green Light. Similarly varieties of apples are being developed that reduce the proportion of small apples grown.

MARKETING

The whole question of marketing may be considered from various aspects.

Grading

Undoubtedly produce must be well-presented to the public if it is to attract the best price. Further, it may be in competition with imported produce that has been properly graded and is free from blemishes. The work involves taking the produce to some special farm building, or to a co-operative grading, or packing, station. The preparation includes washing, grading, packing and sorting vegetables; grading, sorting and packing fruit; and grading, bunching and packing flowers. For these various processes, good light and heating, and in many cases a good water supply, are essential.

When the produce is sorted there will be, invariably, a considerable amount of waste which cannot be sold, but it may be of value on the farm. For example, the waste from root crops and also from brassica crops can be fed to cattle, sheep and pigs, but often this is not done. Pea vines and pods are usually made into silage and fed to dairy cows or to other cattle; but where mobile viners are used, the waste falls to the ground and is then ploughed in to provide humus. When a high proportion of unsaleable waste is expected, sorting should be on, or near, the farm. The use of right varieties, free of pests and diseases, and the harvesting of the crops with minimal mechanical damage, should reduce waste to a minimum.

Special equipment is available for grading onions, brussels sprouts, carrots and parsnips into the sizes needed for various markets but for celery, cauliflowers, cabbage and lettuce hand-grading is essential.

Outlets

After the produce has been graded, it must be sold to the best advantage. This is usually achieved by selling direct to the chain store, supermarket or co-operative society, which generally wants a steady supply of a constant grade. This regularity of supply and grade is difficult for a small farmer to meet, and is more easily organised by a large co-operative group. In future, farmers' co-operatives are likely to play an even more important part in marketing, especially where supermarkets are concerned; they may become even more popular in the future than they are today.

Intelligence

The produce must go to the right place at the right time if it is to make the maximum price. To achieve this, marketing intelligence is most important. This situation is sometimes dealt with by growing on a contract basis. There must be flexibility, for if there is a shortage of a particular product in an area, then arrangements should be made for supplies to be sent. With such crops as tomatoes and plums, which are grown in large quantities in rather limited areas, poor prices may be obtained unless they are sent to an especially selected market.

Display

In order to make produce more attractive to the wholesale purchaser, and in some cases the private purchaser, some roots are displayed in coloured containers: e.g. purple nets for beetroot, white or colourless for parsnips, green for brussels sprouts and cabbage, and red for carrots. For many

TYPES OF FARMING: AIR PHOTOGRAPHS

Plate 1(a)
Fenland farming in Lincolnshire. Note the absence of hedges and trees.
Aerofilms and Aero Pictorial Ltd.

Plate 1(b)
Mixed farming in Kent; fruit trees of all kinds.
Aerofilms and Aero Pictorial Ltd.

Plate 2(a). The coast at Arisaig, Inverness-shire, in the West Highlands of Scotland. The fullest use is made of all land. *Aerofilms and Aero Pictorial Ltd.*

Plate 2(b). Mixed farming in Worcestershire; fruit in foreground. *Aerofilms and Aero Pictorial Ltd.*

OTHER TYPES OF FARMING

Plate 3(a)
Housed laying poultry in batteries.
Farmer and Stockbreeder, Messrs Brian Fender.

Plate 3(b). Difficult farming land near Abergavenny, Monmouthshire, showing the valley productive, but the hills mainly grass. *Farmer and Stockbreeder*, Foster and Impey.

Plate 4(a). Intensive horticulture indoors, in the Lea Valley, Hertfordshire. *Reproduced from 'Agriculture', Vol 61, 1954, by permission of the Controller of H.M.S.O.*

Plate 4(b). Intensive horticulture out of doors, at Marazion, Cornwall. *Reproduced from 'Agriculture', Vol 61, 1954, by permission of the Controller of H.M.S.O.*

HOUSES AND PENS

Plate 5. The system of hurdled folding for sheep. Here Hampshires are shown folded on a root crop. *Farmer and Stockbreeder, Commercial Camera Craft.*

Plate 6(a). Sheep housed on slatted wooden floors. This is an expensive housing. *Farmer and Stockbreeder.*

Plate 6(b). Sheep housed under a plastic roof—a cheaper housing. *Farmer and Stockbreeder.*

Plate 7(a)
Calf rearing in fixed pens. Such calf houses are expensive.
Farmer and Stockbreeder.

Plate 7(b). Sows in indoor stalls. *Farmer and Stockbreeder*.

Plate 8. Indoor beef production: inexpensive building. *Farmer and Stockbreeder*.

MACHINES

Plate 9(a). Track-laying tractor and reversible plough. *Council for Small Industries in Rural Areas.*

Plate 9(b). Farmyard manure: loading for spreading. *Farmer and Stockbreeder.*

Plate 10(a). 18-foot fertiliser distributor.

Plate 10(b). Automatic tractor working in orchard. *Tony Nichols.*

Plate 11(a)
Spinning-disc fertiliser distributor with self-loading scoop. *Mobile Talkies Ltd.*

Plate 11(b)(below)
Self-propelled sugar-beet harvester fitted with top-saver attachment. *John Salmon, Ltd., David Lipson.*

Plate 12. 150-gallon sprayer with 60-foot boom. *Ransomes Sims & Jefferies Ltd.*

fruits the bushel box and half-bushel box are used; these are often returnable but non-returnable containers are now more popular. More recently, partly owing to the difficulty and cost of returning the crates, and also because used crates often become very dirty, the trend is also to use non-returnable crates or containers for lettuce, celery, asparagus and cucumbers. For apples, cell packs have been developed. For many flowers and some fruits, cardboard boxes are used. In these boxes, fruits and flowers can be displayed to advantage and so will often sell at a higher price.

Wrappings

Increasing quantities of fruit, flowers and vegetables are now sold in supermarkets. For this trade, the produce must be clean and wrapped so that it can be mixed with other purchases. Frequently this is achieved most easily if the produce is washed and wrapped in polythene bags or containers. This is not without its disadvantages, for wrapped products must be sold quickly because they have a short shelf life; and condensation, from heat and moisture within the bag, spoils the appearance of the produce. Unless there is a quick turnover after potatoes have been graded and washed and put into polythene bags they will often turn green on exposure to light and have a bitter taste; this is most undesirable; potatoes should neither be put into colourless polythene bags, nor stored in daylight.

Costs of handling

Finally, for growers to receive the best price, produce must pass through the smallest number of hands between grower and consumer. In large co-operative groups, this is achieved by selling direct to retailers, supermarkets and processors, avoiding wholesalers and some of the vegetable markets. Where handling charges and marketing charges are very high, they reduce the prices the grower receives and increase the prices the consumer pays. The total costs of these handling charges and commission in the United Kingdom may reach 15 per cent of the selling price in some markets, whereas in Holland the corresponding figure may be as low as 3–4 per cent.

FRUIT, FLOWERS AND VEGETABLE GROWERS

Space will not permit a detailed discussion of the growing of nursery stock. There are too many facets—providing trees for foresters, hardy shrubs for the topiarist and commercial growers, and the many needs of private householders.

The growing of bulbs is a highly specialised business, especially when the aim is to sell bulbs. The industry is found in counties where soil and climate are suitable, namely the Scilly Isles, in Lincolnshire, and to a lesser extent in Kent, in the Romney Marsh and in parts of the east of Scotland.

Some may think mushroom growing should be included but again this is so highly specialised that it rarely finds its way on to a farm; more frequently in a cellar or darkened barn.

Growers of fruit, flowers and vegetables fall into the following categories:

(a) Specialist glasshouse growers and market gardeners,
(b) Specialist flower growers,
(c) Fruit growers:
   (i) Specialist,
   (ii) Integrated with farming,
(d) Vegetable producers:
   (i) Specialist,
   (ii) Integrated with arable farming.

Specialist glasshouse growers and market gardeners

Growers in this class are far removed from farming since they can control (a) the environment in which they grow their crops, (b) the soil that can be carted in and the nutrients available, and (c) plant pests and diseases by fumigation and spraying.

The equipment to grow the crops is expensive, so it must be used as fully as possible throughout the year. The husbandry demands the employment of a relatively large and skilled labour force.

In general, crops are sold before outdoor crops grown in the United Kingdom are on the market, but sometimes when outdoor crops sent from Europe are on sale in the United Kingdom.

Cloches and frames, both hot and cold, are frequently used in conjunction with glasshouses on a relatively small scale, often in a large old-fashioned vegetable garden but very seldom on a field scale. So this again is well removed from ordinary farming.

Specialist flower growers

These too are specialists who are far removed from farming, since many flowers are grown indoors in very carefully controlled conditions to produce flowers when those grown out-of-doors are not available. Modern techniques have reached such a level that carnations and chrysanthemums can be produced all the year round, whereas naturally they may be available only for a few weeks each year. This is highly specialised work, and so it is for the very special horticulturalist and not for the farmer.

Specialist fruit growers

The production of fruit is normally classified under three main headings:

Hardy fruit that includes apples, pears, plums and cherries.

Bush and cane fruit: raspberries, blackcurrants, red-currants, cultivated blackberries and gooseberries.

Ground fruit: strawberries.

Fruit production may be found on various types of farms, such as specialist farms where fruit is the only crop grown; farms where the top fruit is grown in wide rows with vegetables or flowers planted between the rows of fruit trees; and general farms growing fruit as a sideline on selected fields.

Fruit grown alone. At one time it was common to find mixtures of hardy (top) fruit, bush fruit and ground fruit all growing in the same field, the object being for the strawberries (the ground fruit) to pay the rent for the first 4 or 5 years of the life of the plantation before they were ploughed up, and then for the bush fruit to pay the rent for the next 10 years till grubbed up, and thereafter for the trees to take command, when the land was either bare ground or sown to grassland.

The growing of grass under fruit trees has been a long-established custom; so long indeed that the Kent (or Romney Marsh) sheep have become so acclimatised to grazing under such conditions that heavy stocking of sheep of this breed does not result in an accumulation of internal parasites having any serious deleterious effect on these sheep.

In more recent years it has been considered better policy to keep the grass short by cutting rather than by grazing, as some consider this leads to better quality and colour in the apples grown. The only possible explanation for this seems to be that grazing and the resulting droppings increase the available nitrogen in the soil and so produce the same effect as an excessive nitrogenous top-dressing. Some trees are grown with the land kept bare, or partly bare, by the use of herbicides; this should produce fruit of good colour and quality.

Top fruit trees. After the tree fruit has been established and any bush or ground fruit has been removed, one sometimes finds the land between the trees is cropped with flowers or vegetables; but as will be mentioned later, spraying of trees is then a problem. This policy may be found in Worcestershire, Cambridgeshire and Kent. As a general rule in the space between two adjacent rows of fruit trees only one kind of crop is grown, either flowers or vegetables. They are very rarely intermixed because of the harvesting problems. The vegetables may be cabbages, especially spring cabbages, onions, beetroot or possibly potatoes, while lettuce and marrows may be grown in some orchards. In general these vegetables are cleared before the fruit is ready for picking. With apple trees, especially main-season varieties, there is a long period in which the vegetables may be grown; but if the fruit trees are cherries, and some varieties of plums, then the growing period will be short, for the vegetables must be cleared by late July; but tree spraying, and residues from this treatment, produce problems.

The flowers selected are usually of rather short duration except in the case of Pyrethrums and Scabious. These are perennials and remain in the land for many years; but since they flower in the early spring, no problem arises from their roots being left in the ground during the harvesting of the fruit. Other flowers which may be grown and harvested early are Stocks, and *Chrysanthemum maximum*. Bulbs such as Daffodils, Narcissi, Tulips and also Irises when grown for flowers are harvested before the fruit is ripe; the bulbs will not suffer from being left in the ground and being lightly trampled during the harvesting of the fruit. Gladioli flowers may not be ready for cutting before the fruit is ripe, and consequently these flowers should not be grown between rows of trees.

In apple orchards, with the trees in straight lines, vegetables, flowers and strawberries may be grown on a rotational basis between the rows of trees that will afford protection against light frosts and wind damage. Small tractors can be used between the rows of trees, and rotavators are ideal for the rapid preparation of seed beds for vegetable crops. By growing flowers and vegetables in this way, increased returns are obtained per acre, but sometimes the intercultivation may damage the roots of the fruit trees.

Secondly, intercropping amongst fruit trees will increase labour demands, since greater care must be taken with the extraction of the flowers and vegetables. Thirdly, fruit trees must be sprayed sometimes in the winter, often in the spring and possibly in the summer. There will be little chance of the winter spraying damaging any under-crops, but there may be risks with the spring and summer spraying, so one has to decide whether the crop will suffer more if the spraying is done, or the fruit production if it is omitted. There is also the problem of chemical residues remaining on the fruit.

Fourthly, the colour of top fruit, particularly apples, may be reduced by the manuring of the crops on the land between the trees. It is the nitrogenous manures which, in particular, will tend to produce apples lacking in colour. This may not be serious in cooking apples, but is a great disadvantage with eating apples. The leafy vegetables must have a liberal supply of nitrogen available. In general it is true to say that vegetables and flowers need manuring with phosphates and potash, while the latter is very much needed by top fruit, so that any of these nutrients that may be washed down into the subsoil will be beneficial for top fruit trees.

Fruit-growing integrated with ordinary farming

On ordinary farms where the soil is a deep easy-working loam (light to medium) which is well-drained, and where the climate is not subject to late spring frosts, which may damage the fruit blossom, then it may be possible to plant selected fields with fruit. This usually means a field with a southern slope is chosen, but it is a further advantage if it is well-served with roads and a reliable water supply for spraying. Secondly, some fruit growers consider that a supply of farmyard manure is desirable and this may be cheaper if home-produced than if it is purchased and carted to the farm. Thirdly, it may be necessary to rotate fruit and flowers around the farm, for some growers find this reduces attacks from pests and diseases. Some strawberries have been grown in the same field for 4 or 5 years, and then the field is rested for 12 or 15 years; this has given good results. Such a rotation is more easily achieved on a mixed farm, where there are many different crops, but rotations are difficult on the pure fruit farm.

Blackcurrants will be removed from the site after about 10 years of cropping. With modern mechanical harvesting which cuts off the bushes at almost ground level, it is convenient to grow this crop alone on a field-scale of several acres in extent to avoid moving the harvesting equipment frequently from one site to another. In order to grow clean crops of fruit, spraying is essential; this can be most easily achieved if there is no cropping under the fruit trees.

Fruit-growing makes heavy demands on labour at certain times of the year; sometimes such demands arise at what may be slack times on an ordinary farm, so fruit-growing may dovetail into arable farming. The modern method of storing apples and pears is in refrigerated and gas stores; since these fruits are sold in the winter and spring, labour may be needed to sort and grade them and this may spread labour over the year. On larger farms the gas storage may be on the farm, but smaller farmers may resort to a co-operative store to which they send their labour when it is available. There is also a high demand for labour when planting and pruning, since it is impossible to mechanise these processes; one way in which this situation may be met is by having such work done by contract.

Trees are sprayed at various times during the growing season and this is frequently done, in the case of smaller growers, by a contractor. Picking and carting make big labour demands but here again there is very little possibility of achieving much mechanisation, though the use of pallets and bulk bins is a move in the right direction. Mechanical shakers have been used to shake down apples on to a canvas sheet, but this leads to bruising.

THE PROBLEMS OF FRUIT PRODUCTION

Good yields of fruit will not be obtained from trees that are old and worn out (this may be after 35–50 years). Changed variety demand makes it essential to replace them from time to time. This is expensive and may cost over £100 per acre after grant is paid, and can only be financed by a grower who has a sufficiently large acreage to carry the cost of grubbing and planting, and the resulting lack of income for 5 years, till the new orchard comes into economic production.

The second main problem is that of organising picking, grading and marketing. Since sometimes the small growers cannot do it themselves, they may join co-operatives. This is such an important problem that often even large growers have found it advantageous to sell the produce through co-operatives. The general principle should be to sell top-quality fruit under a trade name to a retailer direct and so reduce costs and increase returns. The best price is usually obtained for a large supply of fruit of uniform quality, regularly available to a big buyer, and this may only be possible with a large co-operative concern. Even large growers may not be able to achieve this. The second quality produce may be sold without a trade mark to wholesalers; but because it is of lower quality it commands a lower price, and since the wholesalers are dealing with it they will need commission. The third quality of produce will go direct to manufacturers either pulped or for pulping, for inclusion in confectionery, pies, etc.

Marketing demands special skill which many farmers may not possess, and also a continuous supply which may be more readily obtained by a co-operative marketing organisation rather than by one individual.

Anyone producing early fruit may find very severe competition from some parts of the Continent where the climate favours growers. This will be particularly the case with strawberries and cherries. On the other hand, the main crop of dessert apples may be little affected, unless competition from France becomes severe. There is even a suggestion that it may be possible to export some of our best eating and cooking apples to the Continent, especially Cox's Orange Pippin and Bramleys. The prospects in general do not call for any undue pessimism.

SPECIALIST VEGETABLE PRODUCERS

Vegetable production in the United Kingdom is found firstly on some holdings where only market garden crops are grown. These holdings are usually relatively small (10–20 acres) and are often near consuming centres.

Secondly, one may find vegetables growing sometimes on a slightly bigger scale in what are known as the market gardening areas of the country, namely the Channel Islands, Marazion in Cornwall, the western coastal areas of Pembrokeshire, around Potton and Sandy in

Bedfordshire, the Evesham district composed of small and large growers, the Lothians, East Perth, Angus, and in the coastal areas of Ayrshire. In these districts it is quite common to find a sequence of vegetable crops being grown, but in other cases, such as in Evesham, in certain areas in Cambridgeshire (the Fens), East Suffolk (around Woodbridge), Norfolk (North and East), on the Lincolnshire silts, in many districts of Kent, and in Perthshire and Angus, single vegetable crops are grown that are followed by ordinary farm crops.

The specialist farms must have a suitable soil and climate such as is found in the coastal strip of East Anglia, where crops grow early and are ready to sell ahead of the produce of the majority of growers. In general the aim is to sow or transplant a sequence of crops. To ensure rapid growth, especially after transplanting, irrigation is employed where possible; for convenience the crops are grown in long strips often as little as one acre in extent. The reason for this "strip system" is that it lends itself to mechanisation and to changing quickly from one crop to another, so possibly obtaining an extra crop for market. The general aim is to have often two crops per year—alternately early potatoes and brussels sprouts as in Bedfordshire—and on intensive holdings three crops per year. This is illustrated by a cropping programme given below:

> First year: early spring cabbage followed by peas, runner beans or leeks (transplanted).
> Second year: brussels sprouts followed by carrots (bunching), onions (bunching), beetroot (bunching), lettuce (boxing) and turnips (bunching).
> Third year: broccoli followed by early potatoes.
> Fourth year: brussels sprouts followed by peas or broad beans (this assumes that time is saved by transplanting the brassicas).

In most of these cases the crops would be grown on chance and not on contract, and they would be sold either to wholesalers or direct to consumers.

INTEGRATION WITH ARABLE FARMING

Turning to farming areas, there are special crops grown in the Channel Islands, especially in Jersey. It is quite common to grow three crops in one season: first early potatoes, followed by transplanted tomatoes and finally transplanted brassicas for stock-feeding. In the coastal areas of Pembrokeshire and Ayrshire early potatoes are often followed by a cereal crop such as barley or oats, or by ryegrass. In the Sandy area of Bedfordshire for many years after early potatoes and brussels sprouts have been transplanted on the same land, and this has been achieved successfully without diseases provided the brussels sprouts are not too late. The pest that one might expect is potato cyst eelworm, but since the crop is lifted early this pest has been avoided. Marazion in Cornwall is famous for the production of early broccoli; after this crop has been removed it is quite common to grow maize or Italian ryegrass.

The single crops are grown traditionally in and around districts such as Evesham, the Fens of East Anglia (extending from near Ely to King's Lynn) and from Lakenheath to Peterborough to the coastal sands of Suffolk (Woodbridge), Holland (Lincs.), and to the slightly heavier soils around Lowestoft and Yarmouth; also more recently in many arable areas of Norfolk, Suffolk, Isle of Ely, Soke of Peterborough, Lincolnshire (Holland) and selected areas in Yorkshire, in the Lothians, Angus and Perthshire in Scotland, and the moss soils of Lancashire around Ormskirk. In these cases, single crops of peas; beans—broad, French and runner; onions; beetroot; carrots; parsnips; celery; brussels sprouts; cauliflower and spring cabbage may be grown integrated with arable crops. Potatoes, especially maincrop, have been handled in this way in many parts of the country for many years. In the Fenland areas rotations are followed, e.g.

First year: Celery,
Second year: Wheat,
Third year: Potatoes,
Fourth year: Wheat.

This keeps the land clear of diseases and pests and gives satisfactory crops. Another entirely different rotation is found in the Fenland and may be as follows:

First year: Potatoes,
Second year: Celery,
Third year: Carrots or parsnips or onions,
Fourth year: Sugar-beet
(This sequence may be varied provided sugar-beet is not grown more frequently than once in four years.)

Here it will be noticed that only one non-vegetable crop is grown, but this sequence also keeps the land free of diseases and pests. Fields are selected for such vegetable growing because they are well-situated near to roads, so that transport is easy, especially in winter time when harvesting may be required.

Vegetables may be taken instead of the normal root shift such as sugar-beet or potatoes, or the vegetable legumes may be peas or beans instead of other legumes, in a cereal rotation. Experience has shown that changes of this nature will improve cereal yields and at the same time give good returns from vegetable crops. When the vegetables are grown in rotation on arable land problems may arise regarding selling. Farmers have found little difficulty in the past in selling potatoes provided they have been properly graded, but they will be wise to make contracts or arrange other market outlets when they are growing such crops as celery, carrots, brussels sprouts, peas and beans. With such contracts the produce is grown for canning, or for quick-freezing. Growers must know exactly what the market requires. For example, the smallest carrots are in demand for canning whole; the very large carrots are in demand for the dicing trade for soups and for dehydration and in some parts of the country for the fresh retail trade; medium-sized carrots are required in the Midlands and in London; larger carrots are bought in the North of England and in Scotland. In order to obtain the best prices the right sizes must be sent to the appropriate markets. This means that the produce must be carefully graded and washed, and all produce showing blemishes must be discarded. This special grading and washing calls for special equipment operated by skilled workers. This is only possible on large farms, or at a co-operative packing or grading centre.

When vegetables are grown under special contract the contractor often makes various conditions. He will supply and charge the farmer for the seed that is to be sown. He will give a definite date for drilling, but the contractor will be responsible for the harvesting, or he will decide on the date for harvesting, or delivery, to his packing station. The growing of brussels sprouts, peas and beans is often on a contract basis, the growing being in the hands of the farmer, but the harvesting is in the hands of the contractor who dovetails processing of vegetables with other activities such as preserving fish. The harvesting of peas and beans demands special harvesting equipment so the contractor usually assumes responsibility.

There is another way in which contracts are sometimes made. This contract system for growing carrots or brussels sprouts may be found in

East Anglia, in Lincolnshire, in Bedfordshire, and in Worcestershire extending into Gloucestershire. In such cases the contractor rents the farmer's field for one season from a certain date by which time the land must have been ploughed and often the artificials applied (frequently as supplied by the contractor). At this juncture the farmer's obligation ceases; the contractor then pays for the use of the field for the one crop, and agrees that the land will be vacated by a certain date to enable the farmer to drill his next crop in good time. The contractor is responsible for drilling a crop such as carrots or drilling and transplanting brussels sprouts. He will be responsible for cleaning the land by use of herbicides and/or cultivations. The contractor will also harvest the crop, grade and market it. The contractor takes all the risks; the farmer is given an assured price for his land. Sometimes the contractor will give a special bonus to the farmer if he is fortunate and has a very good sale, but if he has bad sales he takes the risk entirely himself.

This system of growing vegetables on contract is increasing, and it suits the farmer and the contractor well because:

(1) The farmer has an assured rent that may include an element of profit.

(2) The farmer may be given a special bonus if the crop does well.

(3) There is a change of crop in the rotation without the farmer needing any special skill or any special equipment.

(4) The land may be cleaned of weeds and so may grow cereals, even at a premium, for seed.

(5) This policy is obviously sound crop husbandry.

There is every evidence that the system will develop, because it suits the contractor who is able to grow his crops on "clean" land that has not grown vegetables for several years. He takes his skilled gangs to the various fields to grow and harvest the crop. He will also know when and where it can be marketed to the best advantage.

THE FUTURE OF FRUIT, FLOWER AND VEGETABLE PRODUCTION

The future of production in Great Britain may be affected by trade agreements, but it is relevant to look at the existing situation as set out below.

Table 8.1

Crop	Percentage Home-grown	Notes
Tomatoes	25·8	
Lettuce	44·0	
Cucumbers	58·0	
Celery and leeks	100·0	Signs of increasing celery imports from U.S.A. and Israel out of the normal season
Fresh peas and beans	92·8	
Lettuce	100·0	Signs are that imports of early salads are increasing from Europe
Peas for processing	68·3	
Onions	8·0	
Carrots	75·4	
Cauliflower	87·1	
Cabbage	98·2	
Dessert apples	100·0	Some imported from S. Africa and Australia when homegrown are not available
Pears	34·6	
Cherries	79·2	
Plums	66·7	
Strawberries	97·0	
Raspberries	100·0	Raspberry pulp is imported from time to time
Blackcurrants	100·0	There may be some imports for processing
Gooseberries	100·0	
Other soft fruit—Red and white currants		(Not available)
Other soft fruit, cont.	67·0	Includes blackberries, loganberries, and others, mixed soft fruits, and bilberries and other berries
Cut flowers	81·9 ⎫	
Bulbs	11·5 ⎬ From many European countries	
Nursery stock	18·5 ⎭	

FRUIT, FLOWERS AND VEGETABLES 147

From Table 8.1 it is clear that the production of some vegetables leaves room for considerable expansion, e.g. onions and peas for processing, and carrots, especially for March/June consumption.

There appears to be scope for increased production of tomatoes, lettuce and cucumbers, providing costs can be kept down. The same applies to cut flowers, bulbs and nursery stock.

There is also a case for considering some increased production of dessert apples and pears, and possibly plums.

Competition from the Continent of Europe will undoubtedly be severe, and small growers may be the ones who find greatest difficulty in competing, particularly because of the problem of mechanisation.

The production of main crop vegetables need cause no undue anxiety. It would appear that these can be produced in competition with Europe; in fact, there might be the possibility of some exports to Europe. This may apply particularly to brussels sprouts, peas and beans when sold frozen.

If British growers are to hold their own their produce must be well-grown, well-graded and well-presented to the consuming public.

Note. Further reading may be found in *Dutch Lights for Growers and Gardeners*, by A. R. Carter; *Smallholder Encyclopaedia*, compiled by John Hayhurst; and *Pruning of Apples and Pears by Renewal Methods*, by C. R. Thompson. For details, see List of Books and References on page 240 of this volume.

Various *Ministry Bulletins* are available on many of the individual fruits and vegetables and also on many of the various flowers and bulbs. To these should be added the *List of Approved Products and their Uses for Growers and Farmers* published annually by the Ministry of Agriculture, Fisheries and Food.

Chapter 9

BREEDING OF FARM LIVESTOCK

THE SITUATION TODAY

The British Isles have often been described as the stock farm of the world, and the history of British livestock breeds is already an impressive one when it is appreciated that so many native breeds in many different countries have been improved or upgraded by the use of sires exported from this country. At the present time this trend continues, especially in beef cattle; at the annual Perth Bull Sales the best bulls frequently go to South America and other countries in order to be used on range cattle as a means of increasing beef production. Not only in beef cattle but also in sheep and pigs, the export of British breeds makes a substantial contribution to the nation's overseas trade.

Nevertheless if the most important commercial breeds at present in Great Britain are considered, it must be admitted that most of them are European in origin. In view of the widespread prevalence of the Friesian dairy cow, the Landrace pig, the Charolais beef bull and the recent introduction of various foreign breeds of sheep and pigs, the question must inevitably arise whether the British pedigree breeder has lost some of his old skills, or has merely bred for export and consequently neglected the home market.

In some academic circles there is a current feeling that the breeder has now achieved all that he is able to do, and that if animal productivity is to increase as a result of better livestock, rather than solely through improved methods of management, then some application of genetic theory has become an imperative necessity for the livestock industry. The scientist has already achieved considerable success as a result of his application of "hybridisation" both to the commercial production of maize and other crop plants and to the laying hen. Almost the entire stock of laying birds in this country for intensive units comes from several large-scale commercial breeding enterprises solely devoted to the production of hybrid birds. In the not too distant future, this pattern could well be repeated by the pig industry, resulting in hybrid pigs for pork, bacon, or manufacture being sold under similar conditions to poultry. Hence a decline in the value of pure pedigree breeding, a pronounced interest in European breeds as a means of effecting an increase in the performance of British breeds, and future planning very much orientated towards hybridisation, are the factors most predominant in contemporary animal breeding.

PEDIGREE BREEDERS OF YESTERDAY AND TODAY

The years immediately following the First World War witnessed a considerable interest in pedigree breeds. It was then fashionable to have a pedigree herd or flock; in consequence master breeders were quick to expand their stocks to meet an increasing market for pedigree livestock. This occurred at a time when techniques of management were at a relatively embryonic stage, and it was not long before the late Professor Bobby Boutflour pointed out that, given the correct level of feeding and management, even ordinary non-pedigree dairy cows could easily outyield pedigree contemporaries.

Pedigree breeders themselves are not localised in any particular farming area; they are to be found almost everywhere. This is particularly true of the dairy cattle breeders. Nevertheless there is a tendency for breeders to settle in the more marginal livestock rearing areas of the northern counties rather than, for example, in the eastern arable areas; this tendency is illustrated in particular by the beef breeders. The breeder himself tends to be a man of few words, highly skilled, dedicated alike to the husbandry of his stock and to a lifelong study of the Herd or Stud Book comprising all the pedigrees of his chosen breed. The older type of breeder is often "old-fashioned" in his methods, normally selecting his animals on account of such qualities as "breed characteristics", "scope", "quality" and "balance", all of which are impossible to define but are none the less clearly recognisable when compared with animals without them. The breeder's knowledge of such qualities in his animals has been matched by equal skill in their overall management. Indeed the performance of pedigree stock is often due as much to the standard of husbandry as to any superior inherited characteristics, a point frequently observed by purchasers of such stock.

Show ring standards

In the early years after the Second World War, agricultural shows were as popular in Great Britain as before, awarding prizes for individual animals prepared for, and paraded on, the day of the show. At the same time the Young Farmers' Club movement was promoting stock judging instruction and competition among its young members. These activities had the effect of directing breeders' efforts towards individual animals that displayed these indefinable breeding qualities to the greatest extent.

None the less many breeders soon realised that profit from their livestock enterprises was not wholly dependent upon such characteristics as "a level back", "brightness of eye", or "curve of horn", but rather upon such factors as for example "feed conversion efficiency", "daily rate of livestock gain", or "food consumption per dozen eggs produced". Growing dissatisfaction with show ring results was also felt by the livestock scientist who, disciplined by his training to record, measure and analyse, was actively pursuing a more effective yardstick with which to assess the economic productivity of stock. Primarily as a result of his labours more highly developed systems of recording were introduced, and increasing interest came to be shown in the economic performance of stock as opposed to their physical make and shape.

THE DEVELOPMENT OF ARTIFICIAL INSEMINATION

Side by side with these developments, throughout the 1950's artificial insemination was making rapid progress from its experimental introduction in Cambridge shortly after the end of the Second World War. The use of a superior sire had long been recognised as a time-saving and effective means of improving the quality of livestock, and by employing this new technique a large-scale betterment in the quality of the national dairy herd was envisaged. However, the early years of artificial insemination, while providing the farmer with a cheap and disease-free method of getting his cows in calf, failed to bring about any appreciable genetic increase in the quality of the heifers produced. The principal reason for this was that the Milk Marketing Boards, who had undertaken national responsibility for artificial insemination, continued to purchase their bulls from pedigree breeders who were still using the old yardsticks of selection in their breeding policy. Consequently, in its first ten years artificial insemination was highly successful both from a commercial, as well as from a health, point of view, but genetically it was a disappointment.

Progeny and performance testing

In order to correct this imbalance it became necessary to devise a more comprehensive testing scheme to evaluate the bulls in use. The progeny test, by which the worth of a sire is based upon the performance of his offspring rather than on his own physical characteristics, became the method adopted for this purpose. As a result the Milk Marketing Boards, who still rely on the pedigree breeder to supply the young bulls, now subject their intake of bulls to a highly elaborate testing scheme which takes up to five years to complete before the bull is put into maximum use.

This progeny testing scheme developed by some of the most widely respected livestock geneticists is probably now the most comprehensive of its kind in the world. In addition to evaluation of the animal for yield based on the performance of his daughters, the daughters are themselves subjected to a careful and thorough inspection for those physical characteristics, which have a direct bearing upon their economic life in the dairy herd. Lists of these top rank bulls are regularly published by the Milk Marketing Boards; this enables farmers to nominate these top-class animals for their several requirements, thus achieving a significant increase in the overall improvement of their stock. This testing scheme, together with recent technical developments such as semen dilution and freezing, and the use of liquid nitrogen, has resulted in the establishment of a highly efficient and inexpensive service to the dairy industry of the British Isles.

In the wake of the success achieved by dairy cattle progeny testing combined with artificial insemination, which had so much aided the dairy farmer, the beef and pig producers were also not slow in demanding similar schemes for the improvement of their stock. By this time the Pig Industry Development Authority and the Beef Recording Association had been set up. Both were respectively interested in, and responsible for, an overall improvement in the efficiency of the pig and beef industries. Their first major task was to institute recording schemes from which superior stock might be selected; they were concerned further with sire evaluation schemes primarily intended for the eventual use of superior sires, via artificial insemination, in their industries as a whole. Technically their task was slightly easier than it had been for the Milk Marketing Boards, in that progeny testing of boars is a speedier, less expensive, and simpler scheme to operate than that required for dairy bulls mainly on account of the shorter generation interval involved. In practice the quicker growth rate of the progeny enables a result to be obtained only six months or so from birth. The improvement programme adopted by the Pig Industry Development Authority culminated in the eventual establishment of five boar-progeny testing stations. Between them feeding and management were maintained as constant as possible, so that any difference in progeny performance would be genetic and not the result of management. These stations were well supported by the pig producers themselves, who originally supplied 16 progeny from four different sows to form a testing group.

So far as the breeding of meat animals is concerned, this has always been comparatively easier than the breeding of dairy stock, owing to the well-tried principle that liveweight gain is a characteristic more strongly inherited than milk yield. It is through this daily rate of liveweight gain, an easily measurable factor in the individual animal, that the production of meat is primarily controlled. From this concept has been developed the technique known as Sire Performance Testing. It requires that the individual "meat producing sire" is measured for daily rate of gain during his rearing period, his selection as a future sire being based on his ability to put on flesh at a faster rate than his contemporaries. It was some years before this technique was finally perfected; but now there are several stations in use at the permanent showgrounds of Great Britain where breeders can send young beef bulls. Once they have reached the station, after an initial settling-in period, their growth rate is measured together with their food consumption. The final figure produced is based upon the weight of the animal on a 400-day basis. This weight at 400 days is then taken as a useful measure of a bull's genetic capacity for growth.

This particular technique is much quicker, simpler and considerably less expensive, than full-scale progeny testing. It is currently being operated for both beef bull and boar sire evaluation. Indeed it forms the basis of what is known as the Combined Test for Boars carried out by the pig section of the Meat and Livestock Commission which is now nationally responsible for boar evaluation, and is certainly the most comprehensive and exhaustive testing scheme yet devised for sire selection. Although the older method of assessing sires on the basis of pedigree, coupled with physical characteristics, has had a great improving influence on our livestock over the years, the testing schemes are a significant advance, and should speed up the selection of livestock capable of good performance. The cost of these various testing schemes however is so considerable that organisations, partly financed through a levy system imposed on producers, and partly by the British Government, are required to operate them.

As soon as individual selected sires have been isolated, then through the medium of artificial insemination they can be made readily available to the majority of beef or bacon producers. Artificial insemination, accompanied by its semen dilution and deep freezing techniques, is now securely established and efficiently organised for beef and dairy cattle throughout the United Kingdom. In the pig industry on the other hand, artificial insemination is not yet so well advanced; none the less considerable progress is being made, and it will come as no surprise if within the next few years pig producers are making as much use of artificial insemination as are milk producers today. This at any rate ought to be the logical aim of these testing schemes, ultimately ensuring that the selected superior sires are employed to the maximum extent possible.

All this discussion however, relating to sire evaluation and techniques of artificial insemination, leaves one essential question unsatisfied: that is, who should shoulder the future responsibility for breeding young sires qualified to be selected for these specialised testing schemes? This is a difficult problem, made more difficult by the substantial disagreement which at present exists between scientist and pedigree breeder as to the answer. Some scientists, for instance, would like to think themselves capable of breeding the initial stock, while others even think that breeding is merely a question of putting superior males to superior females. In practice however the problem is by no means so simple.

THE PEDIGREE BREEDERS OF TOMORROW

Earlier in this chapter reference has been made to the pedigree breeder. It is most important that, in a discussion concerning responsibility for future breeding policy, his role is not confused with that of the "master breeder". There is a highly relevant distinction between them.

It must be clearly appreciated that the pedigree breeder is more often than not merely a multiplier of good stock, while the "master breeder" on the other hand has himself actually bred animals of consistent and acknowledged superiority in his own lifetime. It is in the hands of this relatively small group of "master breeders" that the future of animal breeding should lie. Each individual breed may be able to support only a few of them, but it is these men who understand the blood lines which run through the particular pedigree herd book concerned, and who appreciate the necessity of mating together animals of certain lines so that the resultant animals "nick" or blend well together. Further, they realise the necessity of basing the breeding of quality livestock, not only upon the use of superior sires, but upon sound female lines as well. The "tail female", as she is often called, is the most important animal in the pedigree of any bull considered by an intending purchaser.

In order to meet this requirement, the breeder adopts a policy within his herd of always breeding back to some well-proven line; thus all his animals will consequently possess the same "theme" running through their pedigrees. This technique, known as Line Breeding, is only possible to practise if the breeder himself has the necessary specialist knowledge which will enable him to recognise these "good lines", and in addition where they may be located within the breed.

Co-operation between breeder and geneticist

In marked contrast to this approach, the scientist's application to the problem is often more theoretical and mathematical. He tends to over-simplify the matter, seeing it purely in terms of mathematics and a halving process of chromosomes. It would therefore be thoroughly beneficial to the livestock industry as a whole were more active co-operation encouraged between "master breeder" and scientist. Both have an important part to play in the breeding process; first the "master breeder" produces the good animals, and then the scientist screens them by using the best possible techniques. This, in theory at least, ought to ensure that the final selection represents "the cream of the cream". Whether it actually does so or not in practice depends very largely upon the degree of co-operation actually achieved between "master breeder" and scientist. Unfortunately at present there is evidence that this co-operation is lacking in some cases; this can only serve to impede livestock improvement.

Once the final selection of superior animals has been made, what follows? Encouraged by breed societies, the tendency in the past has been for breeders to become set in their methods for the production of "pure breeds". Some, for instance, became ardent supporters of one particular breed, often indeed to the point of intolerance when the merits of other breeds fell to be considered. Although views of this kind are still found today, the barriers to change are finally being slowly dismantled; this is evidenced in particular by a recent trend in favour of systems of planned cross-breeding.

Hybridisation

The phenomenon known as "hybrid vigour" is not in itself a recent development. Whenever animals of different breeds are mated together, it has long been known that the resulting offspring are more "vigorous" in all respects than either of the parent breeds. This practice has been put to profitable commercial use in the breeding of new varieties of crop plants and also hybrid poultry. In the sheep industry, on the same principle, the cross-bred ewe, herself based on a hardy type of hill ewe mated with a "longwool" ram, mated with a "Down" breed of ram for final carcass quality, has for many years provided the basis of fat lamb production in Great Britain. In the pig industry as well, initially impeded by concentration on the pure Landrace as the most suitable pig for the Wiltshire bacon trade, steps are now being taken to reintroduce a planned system of cross-breeding. This system is based on a tough, hardy, coloured breed of saddle-back sow being mated with a Large White boar, the resultant cross-bred sow (or "blue" sow as she is commonly called) then being served with a Landrace to produce a quality pig possessing a high percentage of lean meat. This process constitutes the present foundation of the British pork, bacon and manufacturing pig trades.

The value of cross-breeding has similarly become firmly accepted and demonstrated by the beef industry. For example the famous Blue–Grey (Galloway cow × White Shorthorn bull) is a well-tested instance of an excellent type of hill cow producing calves which, when mated to Aberdeen-Angus bulls (or more recently to Herefords), are ideal for the suckled calf trade.

It is only in the dairy industry that cross-breeding has not been popular. Some years ago British Oil and Cake Mills attempted to produce the Jersian (that is a cross between Jersey and Friesian); but this animal failed to become generally accepted, principally because it was unable to fulfil the dual function of supplying steer calves suitable for rearing for beef, as well as its role as a dairy animal.

The superiority of the cross or hybrid animal over its parent breeds is such that it will not be long before hybrid pigs will be generally available for sale in the same way as hybrid poultry is now. Indeed there are already a few organisations selling such pigs. Likewise the sheep industry is experimenting along similar lines, though its task is more difficult because cross-breeding is already well-established in the industry. Nevertheless different, and superior, cross-bred animals could well come on to the market within the next few years. The sheep industry is particularly anxious to assess the results of the cross-breeding which has continued to make use of the Finnish Landrace breed. In addition to this interest, sheepmen have also recently been turning their attention to the use of French and German breeds as fat lamb producers in competition with the well-established British "Down" breeds.

While the principle of "hybrid vigour" is firmly entrenched in the livestock industry, it is not so well appreciated that the final quality of these "hybrids" ultimately depends upon the quality of the original parent stock. There is therefore a close connection between the old-established closed herd with a Line Breeding policy and the newer ideas of cross-breeding systems, in that the better the quality of the original parent stock the greater will be the quality of the resulting hybrid. In the final analysis the "master breeder" remains a crucial figure in the commercial stock breeding scene. The higher the quality of the stock he can breed, the more exacting the screening process the scientist employs, and the greater the degree of active co-operation achieved between them, the more impressive will be the progress made in animal breeding in the British Isles.

Chapter 10

THE ROLE OF THE PLANT BREEDER

The work of plant breeders has much in common with that of livestock breeders, which was described in the previous chapter. However, improvements can be obtained more rapidly with crops than with stock, since generations can be bred more quickly and discards can be made more cheaply.

INTRODUCTION

Crop production in the British Isles relies heavily upon new varieties for advances in output per acre, and the plant breeder's role is a vital one in this respect. In the earlier part of this century the introduction of new varieties was slower and there were long periods of varietal dominance typified, for example, by the supremacy of Plumage Archer and Spratt Archer amongst the barleys.

However, from about the late 1940's onwards the number of new varieties available from British and foreign sources has increased enormously and, with the royalties available under the Plant Varieties and Rights Act, will undoubtedly continue at great pace. In addition, there has been a great demand for new varieties to produce higher yields in the face of the technical advances made in farming over the last few decades, and as fertiliser use has risen so has the need for cereal varieties that can stand up to high nitrogen applications as well as the newer continuous corn rotations. The spiralling fixed costs of modern arable farming have also increased these demands for more efficient and productive varieties.

It will be appreciated that the work of the plant breeder is one demanding considerable anticipation, for his work is inevitably slow to reach fruition involving, as it does, the necessity to hybridise varieties and raise their progeny to reliable genetic purity before a new variety can be selected. Then must follow critical and detailed field (and in some cases feeding) trials, and finally the on-farm appraisal by critical growers. The task of assessment is large, as is that of ensuring authentic and genetically pure stocks of seed, and in this respect the National Institute of Agricultural Botany in conjunction with the British Seed Trade has come to play a critical and central role. The national trials conducted by the N.I.A.B. and, for example, their Herbage and British Cereal Seed Schemes, are typical of the painstaking and high standards that they have used to ensure that what reaches the farmer is of the very best.

CROP DISEASE PROBLEMS

Before discussing the modern demands within the various crop groups, some general consideration of the disease problem is necessary, though more detail relevant to field conditions has already been given in preceding chapters. Varieties showing increasing resistance to or immunity from particular diseases never seem to last very long in practice, and the reasons for this should be understood.

The disease-causing organisms, in particular the fungi, are not static entities, for, like all biological groups, they are capable of producing a great deal of genetic variation through spontaneous mutation or gene recombination. In this way new strains of the organisms can arise, and these are often referred to as physiological races, which may attack the newer varieties that had hitherto shown resistance.

Examples of this somewhat dynamic field situation, that is, in effect, a response of the disease-causing organisms to natural selection, are all too common, and can lead to a rapid turnover in plant varieties. For example, the winter wheat Rothwell Perdix was affected by a new race of yellow rust and consequently had a short life, whilst amongst the barleys Maris Badger, a good malting variety with high resistance to mildew, showed susceptibility to new races after several years. It is interesting to note that comparable situations are recorded in other groups, and the cereal cyst eelworm is known to exist in several distinct races; whilst in the case of the serious soil-borne wart disease of potato, it is known that a physiological race is found on the Continent of Europe that is capable of attacking British varieties considered to be immune from this problem.

The situation with regard to new varieties is therefore a fluid one, with the demands faced by the plant breeder ranging over a wide field. He is expected not only to improve the physiological efficiency of plants, but to meet the increasingly complex situation caused by the evolution of new races of familiar organisms, and at the same time to maintain quality and be prepared for the sudden ending of a successful product. It will be clear that breeding for what is termed "field resistance", that is for a condition in which the variety is susceptible to a particular disease but is not seriously affected by it, is an attractive proposition; so, of course, is the possibility of producing systemic fungicides and insecticides that could be cheaply applied to control the foliar, stem and root attacks so common in modern field crops. The possible changes in emphasis and methods of crop protection in the future will have a profound influence on the factors required in new varieties. If there is a wide use of systemic fungicides it could result in a shift to more emphasis on the plant's photosynthetic efficiency through breeding for particular habits of growth and more timing of growth in relation to optimum environmental factors than has been done so far.

Cereals

Disease resistance continues to dominate thoughts on improved cereal varieties. All farmers would like immunity or high resistance from the soil-borne diseases like take-all and eyespot. Whilst with the former there appears at present to be little prospect of success, it would be advantageous if the high resistance to eyespot possessed by some wheats could be incorporated in many more new wheat varieties. The desirability of barleys resistant to cyst eelworm (cereal root eelworm), e.g. Sabarlis is obvious for continuous corn growing, and the production of oats resistant to this pest, to which they are at present very susceptible, would lend greater flexibility to cereal rotations. The complexity of disease breeding has already been commented upon but nevertheless mildew resistance in all cereals, leaf blotch resistance in barley and yellow rust resistance in wheat would have high priority in the minds of most forward-looking farmers. In this respect the desirability of continuing long-term disease breeding of a foundation nature such as is being undertaken at the Cambridge Plant Breeding Station is most desirable. Here the complex problem of chromosome transfer involving genes for yellow rust resistance from wild grasses to wheat is a project of major importance.

As the demands for greater yield continue the possibility of breeding for growth patterns must be taken seriously. It is possible that cereals could be produced that would have greater leaf area of greater persistency at maximum light quality periods, more suitable leaf arrangement for light penetration within the crop, and larger ears for more photosynthesis and larger and more grains to act as the "sink" or receptor of increased assimilation.

With rising levels of fertiliser usage, efficiency of nutrient uptake by varieties becomes important and the everlasting demands for the capacity to stand under these treatments will never diminish. Looking ahead there are always numerous general demands. Thus a high-quality malting barley with greater yield, the production of wheats, oats and barleys of widespread adaptability and possessing at the same time a high degree of field tolerance to major disease problems, and varieties that stand ripe and do not shatter, are constantly desired. The possibility of spreading harvest problems by a range of ripening dates is a constant request, but if, say in barleys, it could be a spring variety with highish yield, it would be a major breakthrough. To those who wish for a high-yielding winter barley for this purpose one must suggest that they remember that all winter varieties are, in these days of intensive production, very important in bridging the gap from one season to another as carriers of many important diseases.

With the increasing use of insecticides and fungicides, one future demand is clear. If all new varieties could be screened for resistance to all the commonly used chemicals it would certainly reduce field complications and render easier such widely separated issues as the control of wild oats and leatherjackets.

Hybrid and dwarf forms

Ever since the exploitation of heterosis in the breeding of hybrid maize varieties the prospect of repetition in wheat and barley has been envisaged as a long-term goal. The breeding of hybrid maize varieties has been one of the great biological advances made by plant breeders and involved interfering with the natural breeding system. Maize is an outbreeder or cross-fertilised species in which the male and female flowers are normally in distinctive and separate inflorescences—the "tassel" and the "cob"—of the same plant. In maize the usual procedure has been to produce by self-pollination a series of inbred varieties of poor vigour that will on outcrossing yield hybrid seed capable of producing crops of extra vigour and production. In practice the inbred parents are not vigorous enough to produce enough hybrid seed to make marketing economic and so the double hybrid method was developed to overcome this drawback. The seed has to be continuously produced on the following outline:

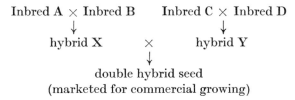

Inbred A × Inbred B   Inbred C × Inbred D
↓   ↓
hybrid X   ×   hybrid Y
↓
double hybrid seed
(marketed for commercial growing)

Unlike maize, wheat and barley are self-pollinating or inbreeding species with small bisexual flowers arranged in dense spikes. If plants are crossed then considerable effort, care and cost is involved in doing so. When varieties of a pure-line nature are crossed they frequently produce hybrids of considerable vigour and potentially greater yield and the use of this heterosis revolves around the production of plants that are capable of natural cross-pollination. In America wheat breeders have developed a complex programme for hybrid seed production. This involves breeding as follows:

(a) for cross-pollinating forms;
(b) for the incorporation of male sterility into one of the parent lines so that it can act as the female parent. A similar line without male sterility is also maintained in order to multiply this line;
(c) for the production of a second parental line that is of normal fertility and is the male parent. This line must contain genes that are capable of restoring the fertility of the seed produced when it pollinates the male-sterile line (b).

The hybrid seed is then produced by suitable arrangement of alternate blocks of both parents with the male sterile being harvested ultimately.

These hybrids have already produced much greater yields than conventional pure-lines and their potential is so great that several British organisations, both public and private, have initiated similar breeding programmes. The problems are great, but if the rewards are achieved then their results will have a dramatic effect on farming in the 1980's.

Hybrid vigour can of course take various forms: it could result in greater tillering or straw length and yield, and careful selection specifically for grain yield will be needed. A timely development has been the introduction of dwarf genes from Japanese and American sources in wheat breeding for semi-dwarf and dwarf varieties capable of standing up to the greater use of nitrogen and on more fertile land. Whilst these varieties are expected in the late 1970's, the incorporation of the genes into the hybrid programme might have an influence on vegetative growth and give the necessary control for high grain yield.

Certainly the future demands in cereal varieties for grain production are numerous, but it is salutary to realise that in many ways they are merely extensions of those that have been requested by previous generations of farmers.

Those farmers who have a use for whole-crop cereal silage will have particular demands. Trials have shown that dry matter yields vary considerably between varieties and according to area, and are not always correlated with those showing highest grain yields. There is a specialised field here and the provision of varieties designed for this purpose is a likely request. Maize comes into the whole crop silage field and will continue to demand some attention, though selection work would undoubtedly be most usefully concentrated into getting grain-producing varieties adapted to the conditions of the eastern and particularly south-eastern counties. This problem involves climatic adaptability and will be a difficult one to overcome. Although rye for winter and spring grazing is of limited use, and is a relatively expensive crop, the selection of more productive and hardy varieties for use over a wide range of field conditions will be a useful addition to the choice of cereal species and varieties available to the farmer.

Herbage plants

With the increasing intensification of grassland husbandry systems, involving as they do the use of very high stocking rates and fertiliser application, the need for more productive herbage plants showing a high capacity to respond to nitrogenous fertiliser and close rotational grazing is obvious. A critical demand in the future will be the possession of a high organic matter digestibility (D-value) so that the maximum nutrients are obtainable from the grazed crop. Digestibility also influences the intake of the grazing animal, and although the restriction on intake tends to occur below an optimum D-value of 63 per cent, there is evidence that very high digestibility does offer high energy availability. It will be natural that breeding for such high values must be essential in the future. Fortunately the development of the *in vitro* technique of digestibility measurement has been of great value. The method simulates rumen digestibility using as little as half a gramme of dry matter, and therefore permits the measurement of individual plants and even parts of plants.

The way is therefore open for the selection of individual plants as possible parents in variety selection. The N.I.A.B. now makes estimations of digestibilities and publishes them as routine in its Recommended Lists; and this important plant character will receive more and more attention from farmers and plant breeders in the future.

The ability of grasses to respond to fertilisers, particularly nitrogen, is known to vary from plant to plant, and so the possibility of selecting for response to the uptake of this and other mineral nutrients exists, and would be desirable.

Plant breeders have been attempting to produce a wider range of growth patterns, and it would be a distinct advantage if an early spring growth could be amalgamated with a medium or better still a late heading time. One of the advantages of Italian ryegrass is its long pre-flowering growth. However this is not present in the perennial grasses (except in the variety S.51 timothy) and it would be highly desirable in an aggressive grass species like perennial ryegrass. The possibility of obtaining a perennial variety capable of producing early bite would be of great value too and give much lower costs of production compared with the short-lived Italian ryegrass normally used. Whilst tall fescue comes into the early-bite category, its lower digestibility and coarse nature has not made it attractive to farmers, though breeders have been attempting to develop fertile polyploid derivatives from crosses between Italian ryegrass and tall fescue in the hope of combining the quality of one with the persistency of the other.

The tetraploid ryegrasses have shown that it is possible to produce alterations in the biochemical constituents of varieties, and in their case an increase in soluble carbohydrates is probably responsible for their greater palatability. It could be advantageous if new varieties could show some specialisation in certain directions, for say high carbohydrate as opposed to high protein composition, so that they could be used in particular ways and possibly supplemented by balancer feeds.

The possibilities of producing non-flowering varieties continues to be an attractive one as it would reduce management problems. Grass breeders have been examining the feasibility of producing varieties with greater photoperiodic requirements for flowering than can be obtained in this country and producing seed in more northerly latitudes where day-length is longer. One should remember that the approach of selecting plants capable of trapping solar radiation more effectively than others would be desirable, and might lead to a lowering of the recovery period used in rotational systems to as little as two weeks. Before this could be effectively used in conjunction with high nitrogen applications, it would be necessary to select against nitrate presence in grasses following such treatment, as this may be harmful to animals eating the herbage. Then again, the need to select varieties with lower incidence of foliar diseases which tend to lower feeding value would be of great advantage.

In herbage legumes a great deal seems possible, since these plants are often, as is white clover, of greater digestibility than the grasses, as well as being rich in proteins and minerals. The production of a white clover capable of fair survival in heavily fertilised pastures would be a real advance, and would go a long way to improving the meat-producing capacities of intensively managed swards. From all these possibilities, it is clear that there is an increasing enterprise in the contribution of geneticists to British farming, and also that a new demand is awakening among farmers in the realms of grassland production.

Root Crops and Break Crops

The future demands in root crops will be rather specialised; this group is likely to face increasingly the pressures of rising costs of production. In sugar-beet there is need for more varieties of the monogerm type that will permit precision drilling to a stand though they will need to possess the high standard of field characters and performance of existing multi-germ types. Sugar-beets showing a high degree of resistance to bolting, even varieties capable of earlier drilling without bolting, will be possible attainments. In this specialised crop, disease problems such as virus yellows and downy mildew are ones that farmers would like resistance to. In the less specialised kale crops, improvements in digestibility, leaf-to-stem ratio and winter hardiness would be advantageous. Increased yield will be desirable, and if goitrogenic properties can be eliminated from all such productive stocks it would be a great advantage.

Although the swede crop is now of minor importance, its value especially to certain marginal areas still merits attention, and one must hope that some of the wide range hybridisation work being investigated might lead to higher levels of production and usefulness.

Growers would like to see potato varieties with a range of improvements. Resistance to potato blight and virus diseases has proved a difficult task, but the demands for such features will continue, as indeed will the requests for varieties that are clear of the numerous skin blemishes, such as common scab and skin spot, that affect the tuber. The production of potato seed stocks from stem-cuttings is a new and most promising development in producing disease-free stock. Perhaps the greatest achievement would be varieties resistant to root eelworm, and farmers and processors will continue to request improvements in specialised aspects of quality.

Modern break crops such as beans and oil-seed rape have shown some weaknesses. In the former, greater resistance to chocolate spot and the developing aschochyta disease problem would be very useful improvements. Farmers would also welcome a variety that was vigorous coupled to high self fertility in order to obtain an assured seed set and avoid dependence on pollinating bees. Oil-seed rapes showing more even ripening and less tendency to shatter would be desirable achievements.

PLANT BREEDING AND HUSBANDRY

Finally one can turn to the future influences that plant breeding may have on husbandry methods. Clearly the future economic prospects of the British farmer depend to a great extent on his ability to utilise in full the raw materials of the physical environment. Successful plant breeding plays a vital role in achieving this objective, for in producing varieties more efficient in photosynthesis, mineral uptake and disease resistance, their output potentials are raised. However, these plants require good field conditions for realising their yields, and it is clear from recent trends, particularly continuous corn growing, that the environment is far from the optimum desired. Thus in this system the build-up of the soil-borne organisms, take-all, root cyst eelworm and eyespot, seriously reduces yields by so affecting the root, stem base or vascular tissues as to hinder the uptake of soil water and nutrients and impede their movement in the plant. Although these cropping methods lead to lower yields, they can have economic success through low costs of production, and therefore in such an environment it is resistance to these troubles that would be the major factor leading to higher yields. Whilst good eyespot resistance is present in some wheats, it is absent in barley, as is take-all resistance in both species.

It is a modern paradox that some crop husbandry practices limit the value of newer high-yielding varieties by creating environments containing what appear to be insoluble problems, problems that are of small effect on farms that practice alternate husbandry. Obviously the plant breeder cannot dissociate himself from newer practices, and whilst he must be able to appraise their demands, it is not implicit that he will find the genetic material with which to meet them. In continuous cereal growing it is possible that the future will rest more with the development of potent soil fungicides. However, the move to seed-producing break crops within the cereal sequence does offer the plant breeder the opportunity of direct aid, by breeding higher yields and reliability of performance into crops such as beans, peas and oil-seed rape.

If in some ways the future prospect appears unfavourable for the intensive cereal farmer as far as help from the plant breeder is concerned, it is somewhat more encouraging for the grassland farmer. Here increasing livestock output will eventually depend on increased herbage production and quality, and plant breeders now have the ability to measure these factors in detail. The trends in selecting herbage plants for higher digestibility and mineral uptake, as well as appreciating the need to produce genetic types capable of responding to intensive grazing that at the same time are capable of utilising solar radiation efficiently, hold considerable promise. Even for zero grazing, it is likely that selection of types more suited to cutting treatments can give greater potential to the system. Then again, in all livestock grassland enterprises the possibility of selecting for variability in feeding qualities in factors like carbohydrate or protein values remains largely unexplored; but if developed, this may provide new livestock feeding methods with emphasis on energy requirements from grassland and protein from elsewhere.

Thus the plant breeder has not only to wrestle with problems arising from developing husbandry systems, but must of necessity be forward-looking. He must search constantly for new genetic diversity within species, which may answer problems and also create new avenues of exploitation by the farmer. The effort which goes into long-term hybrid cereal breeding epitomises his approach. However, the task is not his alone but must be shared by a responsible and realistic husbandry approach if the high yields possible in modern crops are to be properly exploited. It is no use breeding a more productive plant machine if habitat optima are lowered by other factors. The farm is a complex environment, and requires the joint approach of many disciplines to give the best results.

The ranges of possible improvements desired within the numerous crops grown by the British farmer are many. They represent not only an increasing awareness of the limitations of crop plants as "photosynthetic machines", but also a greater emphasis on plant health and field performance, which has developed in crop husbandry over the last few decades: an emphasis that will intensify as economic pressures and technical education continue.

Indeed, plant breeders must forecast the needs of the industry 15–20 years in advance. With constant rise in labour costs, one of their preoccupations must be the effect of increasing mechanisation upon the crops he produces. Various aspects of farm mechanisation are considered in Chapter 11.

Chapter 11

FARM MECHANISATION

The mechanisation of British agriculture is a development which has progressed steadily over many years. Over the last 30 years or so progress has been speeding up; over the past ten years the rate of development has been very rapid indeed, with a great many sophisticated tractors and machines being used to perform operations better and quicker than ever before. Automatic control systems are now fitted to certain items of fixed equipment, and these help to reduce their labour requirements. In most cases the limits for further progress at present would seem to be economic rather than technological.

The largest problem of economic mechanisation in the United Kingdom is the number of small farms. It is easier to achieve economic mechanisation on a large enterprise since the equipment can be more fully utilised. It is impossible to scale down equipment in most cases, and the result of this is that the cost of mechanisation tends to be higher on smaller farms, and greater use has to be made of contractors' services, syndicate ownership of machinery or the purchase of second-hand machinery to keep the figure within economic limits.

Syndicate ownership of equipment shows great promise as a means of reducing mechanisation costs; although many farmers seem a little wary of it, tending to look for difficulties rather than advantages. There are quite a number of schemes in operation covering a very wide range of equipment and not all limited to small farms.

It is difficult to compare capital investment per acre on farms which are not similar in all major respects; but the figure is likely to be in the region of £22 per acre on farms of approx. 100 acres to £15 on farms 750 acres (Ref. 1).

THE OBJECTS OF MECHANISATION

Five objectives of mechanisation can be stated, as follows:

Increase in profit,
Reduction in labour,
Elimination of manual work,
Timeliness and improved husbandry,
Higher output and better quality.

These factors are inter-related, and although increase in profit is the main objective this should not be considered in isolation. A reduction in labour does not necessarily mean a reduction in overall cost. Saving part of a man's time is of no value unless he is going to be gainfully employed elsewhere on the farm. A reduction in the need for casual labour or a levelling-out of peak labour requirements has far more value, for example with such operations as potato planting and harvesting, and sugar-beet thinning. A reduction or elimination of heavy manual work is desirable from the point of view of improving working conditions. It may not always be an economic proposition to mechanise an operation, but if it leads to better working conditions and hence to a contented farm staff it may well be worth a slight loss of profit.

The effects of timeliness and better husbandry can be easily seen and appreciated in terms of higher crop yields and better quality. The ability to perform an operation at the correct time, and in the correct manner can be all-important. Only a certain number of days during the year are available for particular operations, and it is essential that the machinery capacity is such that the operation is completed in the time available. One of the greatest benefits of increased power of tractors and size of equipment is an improvement in timeliness. It must however be realised that farm mechanisation is only a tool of farm management, and unless the overall management relates the cropping, stocking and mechanisation in a sensible way, little will be achieved.

TRACTORS

At the present time the range of tractors available is greater than ever before. The choice of types, horse-powers and optional fittings is such that practically all situations can be provided for. Generally the power output of tractors has shown a significant increase over the past few years, and although the number of tractors in use has tended to stabilise, the amount of available power has increased. Excluding market garden tractors (less than 10 bhp) and tool frame types, the range of sizes now available is between 25 and 160 bhp. Table 11.1 shows the range of types, sizes and approximate prices of tractors at present in use.

Table 11.1
Details of tractors

Type	bhp Range	Size group	Price range £/Tractor	£/bhp
	Less than 25	Small	No longer available in U.K.	
Wheeled 2-wheel drive	25–44	Small/medium	770–1200	30–32
	45–59	Medium	1210–1400	26–29
	60–75	Large/medium	1440–1860	22–26
	76–99	Large	2665–3185	28–35
	over 100	Very large	3400–5600	30–39
Wheeled 4-wheel drive	45–50	Medium	1500–2390	27–43
	60–75	Large/medium	2300–2780	35–42
	76–99	Large	3215–3905	37–42
	over 100	Very large	3420–6154	32–40
Tracklayer	45–60	Small/medium	1990–3910	40–65
	61–80	Medium	2850–5025	46–68
	over 80	Large	3700–9625	41–75

It will be seen that in terms of cost/bhp, small tractors are expensive. This is because certain items of production costs are similar whether the horse-power is large or small or whether many or few are produced. Since the engine size is small, the cost per bhp is high. The medium-powered groups appear as the best value for money. This is due to the numbers produced allowing the benefits of mass production to be obtained on a larger power output. The medium (45–59 bhp) group is the most popular size at present; the large/medium group are higher in first cost but cheaper per horse-power, and are becoming used more as labour costs increase.

Very large-wheeled tractors are found on large farms where their power can be justified. There is no doubt that these tractors are expensive in terms of both purchase and running costs, and if their potential is to be utilised they must also be equipped with expensive implements and machines capable of withstanding the stresses set up by the high power and high speed of working. The cost of operating one of these tractors becomes excessive below about 700 hours of full load operation per year, but if one very large tractor is used to replace two medium tractors and the farm is run with one man less, then their use can be of great value. In practice this is however not always achieved, and perhaps the greatest value of this type of tractor is its ability to work at very high rates when conditions are suitable, so maintaining timeliness with some reduction in labour requirement at peak periods.

The 4-wheel drive tractors produced in the United Kingdom incorporate some parts designed for tractors in the large/medium group. This allows some benefits of mass production to be gained and keeps down the total cost. Four-wheel drive tractors perform better under heavy conditions where wheel slip is a problem, and on hill sides where stability and traction are problems. The type of tractor with all 4 drive wheels of equal size gives the best performance, particularly in the larger sizes, and there is a definite increase in the number of this type of tractor.

Tracklaying tractors are expensive, owing to the way in which they are constructed and the fact that they are produced in relatively small numbers. Some farmers on heavy land are concerned about the possibility of soil compaction from the use of very heavy-wheeled tractors, and consider that tracklayers are necessary to prevent damage to the soil structure. By virtue of their construction, this type of tractor exerts a much lower ground pressure* than wheeled tractors; they do not have the same tendency to cause rutting and smearing of the soil. Their disadvantages are well known and often overstated, namely high cost,

* The ground pressure of a tractor is its weight divided by its area of contact with the ground.

lack of speed, and difficulty of moving on roads. The use of hydraulically operated 3-point linkage systems designed to enable semi-mounted and fully-mounted equipment to be used on tracklayers increases their versatility and ability. The use of better materials for track components also means that track replacement can compare quite favourably with tyre replacement costs on very large-wheeled tractors on some soils. The slow forward speed and difficulty of moving from one part of the farm to another where roads have to be travelled will remain disadvantages, but in spite of this the numbers of tracklayers on farms has shown a slight increase recently, and interest in the medium and large groups, as opposed to the small types, is increasing.

Number and type of tractors required

It will have been realised that a "tractor" is a very variable piece of equipment so far as size and ability is concerned. The concept of "standard tractor hours" was put forward by Sturrock (Ref. 2) as a means of estimating the number of tractors required for a particular method of farming. The size of tractor taken as the standard was the most popular size at the time this exercise was carried out, this was approximately 40 horse-power and assumed to be capable of pulling 3 furrows. A small tractor capable of pulling 2 furrows was treated as 0·65 of a tractor, and any tractor capable of pulling 4 furrows as 1·3 tractors. At the present time a 40 horse-power tractor would be regarded as a small/medium size (Table 11.1), and it is obvious that some extension of the original concept is required to enable it to be used on the larger tractors now available. Full utilisation becomes more difficult as horse-power is increased, and doubling the horse-power does not double the rate of work of the tractor over the whole year. Table 11.2 shows estimated ratings for the present range of tractors.

The number of tractors required for a given farm can be estimated from figures of hours of work required annually for the crops and stock in the farm programme. It is advisable to use figures obtained locally for this purpose, since they will be a more accurate reflection of conditions; but tables of requirements are available in publications on farm management (Ref. 3), and can be used as a guide if no others are available. The first 800 hours of tractor work per annum can be completed by a single tractor. After this number of hours, a second tractor is required due to the fact that it becomes necessary at times to have two separate operations going on at the same time, possibly on different parts of the farm. The second and any further tractors can

FARM MECHANISATION

Table 11.2
Rating of tractors

Group	No. of furrows	Rating (Std. tractors)
Small	1–2	0·65
Small/medium	2–3	1·00
Medium	3–4	1·25
Large/medium	4–5	1·65
Large	5–6	2·00
Very large	6–8	2·25—2·50

each be expected to complete 1,200 hours of work per annum on average, although this will vary according to local conditions and type of farming. By a study of the farm programme an estimate of standard tractor hours can be made, and any seasonal peak requirements should become evident by relating the programme to the average number of days available for specific operations. These peak requirements are more important than average annual requirements since they can affect timeliness. The study will also show the amount of various operations required, e.g. ploughing, rotary cultivation, forage harvesting. From this information coupled with a knowledge of soil conditions, gradients etc., the size, type and number of tractors required can be estimated. The final choice of make of tractor should be made with due regard to the availability of good service from a local dealer.

TILLAGE EQUIPMENT

Ploughs

The advent of larger tractors has led to a number of developments in basic cultivation equipment in the United Kingdom. The plough is still the most important basic tool, but there are some difficulties in matching the increased power to the equipment available. High-speed mouldboards have been developed for ploughs to enable better use to be made of higher working speeds, but there is some resistance on the part of many farmers and tractor drivers to the use of these higher speeds. High-speed work is more successful on light soils, but it is possible to increase the speed from the often-used 2·5 up to 4·0 mph with modern ploughs on heavy soils without deterioration in the finished work.

The most popular type of plough at present is the fully mounted type. These ploughs are usually restricted to 5 furrows owing to their effect on tractor stability when out of work, and semi-mounted types are popular for use with the larger tractors. Reversible ploughs are being used increasingly. They have the advantages of higher rates of work owing to less marking-out and setting-up, and leaving work free from ridges and open furrows. This latter factor is important where a level seed-bed is required for a crop such as sugar-beet, and is also desirable where large combine harvesters or forage harvesters are to be used. The disadvantage of reversible ploughs is their increased cost, which can be nearly double that of the equivalent width orthodox implement.

CULTIVATION EQUIPMENT

A wide variety of implements and machines is available. Their uses vary from deep work, e.g. subsoiling and breaking up soil at shallower depths, to preparing a tilth for seed-beds and surface cultivations to destroy weeds.

Cultivation equipment is generally used after ploughing, but some farmers are now using "chisel ploughs" which are essentially a robust form of tined cultivator to replace mouldboard ploughs. Working to the same depth as conventional ploughs, a comparable finish can be achieved in 2 or 3 passes, at a cost similar to ploughing.

Many heavy cultivators have been developed to harness the higher horse-power now available from modern tractors. A feature of most of them is the ease by which tines or even whole sections of the implement can be added or removed to allow use on tractors of different powers or for working at different depths with the same available power. The types of tines fitted vary. Some are heavy spring steel; others fixed, solid or fabricated steel. Various tine angles are used, and whilst claims are made by some manufacturers about the greater effectiveness of their particular shape of tine, in practice the margin for difference is quite wide, except in the case of spring tines where the pull on the implement can alter the tine angle in work and so reduce its penetration ability.

Lighter types of cultivator are used for shallower work on seed-beds.

The most popular implement at present is the spring tine harrow/ cultivator. Worked at high speed the tines vibrate and have a very good shattering effect on the soil. For cereal seed-beds in spring, quite often one pass on a frost mould is sufficient after winter ploughing to produce a seed-bed. They will work to a depth of approximately 6 inches and are available in widths up to 20 ft for various sizes of tractor.

"Harrows" is a term which covers a remarkable variety of implements used for seed-bed preparation, seed covering, weed destruction and on grassland for aeration. Zig-zag harrows are used in seed-bed work, a variety of weights and tooth lengths being available to suit different soil conditions. The present trend is towards attaching the harrows to a frame mounted on the 3-point linkage of the tractor to facilitate handling and transport. Heavy wooden-framed harrows known as "Dutch" harrows are used by many farmers for sugar-beet seed-bed preparation. This harrow acts as a leveller and its straight tines break down small clods without bringing unweathered soil to the surface.

Flexible harrows are made from a mattress of metal links which end in spikes which penetrate the ground. Some flexible harrows have adjustable knives fixed into blocks attached to the links. This type is heavier and much more effective in aerating grassland.

Disc harrows are available in a range of weights and widths, either trailed or mounted and in tandem or off-set forms. The trend over the past few years has been towards heavier trailed types to utilise the power available from larger tractors. They are excellent implements for preparing a seed-bed after ploughing, since they do not bring rubbish to the surface and they have a firming effect on the seed-bed. Some farmers use heavy discs prior to ploughing-in stubble to assist in burying straw and rubbish and to control grass type weeds.

Power-driven cultivators are being used in increasing numbers at present. The increase in available horse-power and the better speed ranges of modern tractors have done much to encourage the use of the rotary hoe type of cultivator. The main advantages of this machine are its ability to produce a seed-bed in the minimum number of passes, so preventing loss of moisture, or to force a tilth in a difficult year, cutting up and burying surface growth and trash. Correctly used it can also control such weeds as couch grass. Its disadvantages are that it has a high power requirement, the initial cost is fairly high, and maintenance and repair costs can be heavy.

The Howard Rotavator Company have developed the "Rotacaster" which is an attachment fitted to their rotary cultivator, enabling cultivation and drilling to be done in one pass on soil which has had only one cultivation since the previous crop. The same company have also developed the "Rota-seeder" which is intended for use in minimum cultivation techniques following the use of chemicals such as Paraquat on grassland and crop stubbles. This machine merely cuts slits in the soil into which the seed is dropped. The actual amount of ground disturbed is only about 5 per cent of that disturbed in normal cultivation.

Another form of powered cultivator is the reciprocating harrow. This is a series of tines mounted on bars which are caused to reciprocate by a drive from the power take-off on the tractor. The advantage of this machine is its ability to produce a tilth to the full depth of its tines without bringing unweathered soil to the surface and to produce a seedbed from weathered ploughing in one pass given the right conditions, so saving time and reducing the risk of compaction. Its disadvantage is its fairly high initial cost.

A farmer will have a selection of cultivating implements at his disposal, the actual sizes and types being dependent on the tractors he uses, the type of soil and its condition, and the crops being grown. The type of implement used and the timing of its use are much more important on the heavier types of soil. Cultivation implements are relatively inexpensive to purchase and have a long life; the cost of using them is therefore fairly cheap. Minimum cultivation techniques using chemical sprays are not necessarily cheaper than using cultivations, and in some cases may be more expensive. There are circumstances, due to weather or soil condition, where cultivation cannot be done effectively and spraying provides a useful alternative method of weed control. If slit seeding is used for a crop such as kale, weed control is maintained until the crop is established; this reduces the need for inter-row cultivation. Also the presence of the old turf reduces poaching of the soil surface by the animals grazing off the crop.

FERTILISER HANDLING AND SPREADING

Many developments and improvements have taken place in fertiliser spreading machines over the last few years, but a great deal of attention is now being paid to improved methods of handling the fertiliser as a means of reducing the cost of spreading. As with all farm materials, unnecessary handling only adds to the total cost of the operation; better organisation of storage, handling and spreading can lead to substantial increases in rates of work output per man and a possible reduction in the overall cost of the operation.

A high rate of work is essential so that other work is not held up at peak times. A high output per man is essential since the farm staff will be fully committed at these times. In spite of this, some farmers consider it desirable to have a second man to help in the operation of opening bags and filling the hopper to speed up the operation. The cost is important but must be weighed against such factors as timeliness, convenience and reduction of heavy manual labour.

On many British farms the time involved in handling fertiliser from the storage point to the spreader is much more than on the actual spreading operation. It also may involve a lot of heavy manual work. In these cases great improvements are possible through better methods of handling.

Handling in sacks has been the accepted method of handling fertiliser for a long time. It does have several undesirable points concerned with intermediate handling and storage, and can never be as efficient as a well-designed and organised bulk handling system. Nevertheless efficient sack handling systems are possible and are considered more desirable than bulk handling by most farmers. Reduction of manual handling, and all unnecessary handling, can show big improvements in output. Plastic sacks can be stacked without the necessity for completely water-tight buildings, and they can be stockpiled in the open on sites near where the fertiliser is to be spread. Stacking on platforms made from old railway sleepers or old lorry bodies at hopper or trailer height cuts out lifting and speeds up the operation of outloading.

Taking delivery direct on to large trailers gives the advantage of being able to move the fertiliser along to keep up with the spreader as it proceeds across the field. Handling on pallets can cut out manual work, but this system tends to be expensive and is perhaps better suited to larger farms where the handling equipment can be more fully utilised on other operations. Handling in bulk enables better output to be achieved from one-man systems. It is however better suited to farms where the annual usage of fertiliser is in excess of 50 tons and where only one or two different grades are used. There are difficulties in keeping proper control of sowing rates, and storage stocks, but these are not insurmountable with good management.

If bulk fertiliser is to be stored, this will increase overall costs; but elaborate self-emptying containers are unnecessary, and floor storage in a dry building need not be very expensive. In some areas there is a reduction in price for bulk fertiliser and on large amounts the reduction in cost through buying in bulk and out of season can offset the cost of storage.

The most effective method of handling bulk fertiliser is direct into the hopper of a large spinning disc-type spreader at the store, using a tractor scoop. Where this is not possible due to the type of spreader being used, or when combine drilling, the fertiliser can be loaded into a tipping trailer fitted with a spout in the tail board and used in conjunction with portable ramps so that it can tip direct into the spreader. A more recent development which shows much potential is the high lift type of tipping trailer which can either tip direct or via flexible trunking into the spreader.

Methods of spreading

There are basically two types of spreader: those in which the spread width is equal to the width of the hopper, and those which broadcast the fertiliser over a width considerably wider than the machine itself. A recent introduction is a pneumatic type spreader, which employs a narrow hopper but has booms similar to a sprayer for actually spreading the fertiliser. This combines most of the good points of the other two types.

Full-width machines will spread evenly across their width without any skill from the operator; accurate joins between points can be easily made. They tend to be expensive and are slow compared with the broadcasting type. They are more difficult to transport, the larger versions having to be either pulled from one end or loaded on to a transporter. Because of this they do not lend themselves to simple bulk handling systems.

The non-fixed width machines are available in small or large hopper versions. They are relatively cheap and capable of operating at greater speeds than full-width types. The large types carry loads of 25 cwt or more, and since transport width is the same as working width they can travel through gateways and on roads, enabling them to fit into bulk handling systems very well.

The main criticism of these machines is the possibility of unevenness in spreading. Machines available at present are very much improved over older versions and with intelligent machine settings and bout overlapping a satisfactory performance can be achieved.

CEREAL DRILLING

At present a great deal of thought surrounds the operation of drilling cereal seeds in Great Britain. Whether it is autumn or spring work, it creates a critical period when all efforts are turned towards getting the crop in at the optimum conditions. The typical machine used has been a combine grain/fertiliser drill with 15 coulters at 7-inch spacings. This type of drill generally has an output of approximately 2·5 acres per hour with 2 men, although better organisation and methods of handling can result in outputs of up to 3·5 acres per hour.

Drill design, particularly hopper capacities which are too small and coulters which demand low speeds, restrict further increases in output. A medium large tractor can handle a much wider drill than is normally used, and farmers are now turning to larger drills or using two drills coupled in staggered tandem.

Dividing the operation into two parts is another method being practised. The fertiliser is broadcast with a high output machine and the drill is used for seed only. The reason for this is that it is usually the fertiliser hopper of the drill which is too small in capacity and hence requires more frequent filling. A plain grain drill with a large hopper can show significant increases of output over a combine drill, but many farmers still believe that there is a good case for combine drilling. In general it can be expected that combine drilling will give an increase grain yield of about 1 cwt per acre or more over plain drilling on soil of low fertility. A loss of 1 cwt per acre could easily be made however, through lack of timeliness due to protracted drilling.

Most plain drills have coulter spacings of $4\frac{1}{2}$ inches to 5 inches. This spacing can result in yield increases of $\frac{1}{2}$ cwt per acre compared with 7-inch spacings, and where narrow rows are practicable a case can be made for plain drilling. On wet heavy soils, narrow coulter drills are liable to difficulty due to blocking and increased draught; best results are obtained on fertile lighter soils. Another approach is to broadcast both seed and fertiliser together by means of a fertiliser spreader, and then harrow in the seed. This can make savings in both time and labour, and some farmers claim an output of 100 acres per day with no significant differences in yield compared to drilling. This operation requires good soil and a good seed-bed, but sowing can often take place earlier due to the lower weight of the tractor and spreader compared to a larger tractor and drill. Mixing can be done by tipping on the floor and handling with a tractor scoop, final mixing taking place in the broadcaster hopper. In one instance a farmer has purchased a second-hand "ready-mixed" concrete truck; this holds sufficient material for a full day's drilling and acts as transporter, mixer and hopper filler in the field.

Direct seeding into stubbles killed by chemical sprays shows possibilities over a range of soil conditions; very robust and heavy drills designed for this purpose are now being produced. Special coulters are used to obtain penetration.

CROP SPRAYING

Spraying machines are used to apply a wide variety of herbicides, insecticides and fungicides, and much progress has been made in the design of sprayers and in the use of non-corrosive materials in their construction. Simple "low-volume" types capable of applying 5–20 gal per acre are quite cheap, and although there are limitations to the materials such a machine can handle, the range is adequate for weed control on cereals. Where greater versatility is required, more expensive "universal" machines are available. Probably the greatest advance in design of sprayers is the use of wider boom widths. This not only increases work output from the machine, but also reduces the amount of damage done to the crop. Booms of up to 60 ft are in use, and these wide booms are usually mounted on a hydraulic fore-end loader mounted on the tractor. The spray tank is mounted or trailed at the rear of the tractor. If row crop type wheels are used on the spraying tractor, the damage to the crop is reduced even more. Whilst it is possible for a farmer to buy and operate a sprayer quite cheaply there is a considerable amount of contract spraying done. The main reason for this is the increasing technicality of applying some of the newer sprays, with risks of damage to crops and injury to staff who are unskilled in the use of potentially dangerous chemicals.

CEREAL HARVESTING

Although a few binders are still in use (especially on small farms in the west of the U.K.) the acreage handled by this method is of little consequence and is reducing annually. The total man hours per acre involved in combine harvesting are only about one third of those involved in the binder method, and a much larger acreage can be dealt with in a short time by a small team. Most combines in use now are of the self-propelled type. The range of outputs is between 2 and 16 tons of grain per hour. Cutting widths between 6 ft and 18 ft are available, the wider cutting tables being easily detached and towed behind the combine for transport. Most grain is now handled in bulk and not in sacks. It is usual to rate combine capacities in terms of tons per hour rather than cutting widths, since there is no definite relationship between cutting width and output. In fact many models are made with alternative table widths, e.g. 8 ft, 10 ft, or 12 ft, all the other parts of the combine being identical. The reason for this variation is to cater for different crop yields and amounts of straw which might be found under different conditions.

It is essential when looking at manufacturer's, or official, test figures for outputs that "net" outputs are not taken as the ability of the combine. These figures relate to when the combine is working continuously at a set loss of grain per acre. Under actual farm conditions, the output is not likely to be much better than half this figure, and in laid or wet conditions even less. The size of combine required for a given acreage is affected by several factors but the choice will normally end up as a compromise between the ability to clear the crop in good time and the amount of capital which can be invested.

The number of hours available for combining will have a great bearing on capacity. This is a very variable factor and may lie between 150 and 200 hr in the eastern parts of Great Britain, and rather less in the more northern and western parts. It is not economic to buy a combine capable of dealing with the worst possible harvest, but as much capacity as can be afforded should be available to reduce the risks in a bad harvest. If average figures are used for this assessment the farmer is likely to be in difficulty every other year; in general, 25 per cent more capacity than would be needed in an average year is to be desired as a minimum. This applies to all other equipment allied to the grain harvest and not just to combines.

Having a good margin of capacity will enable the farmer to limit combining operations to when the grain is at a reasonable moisture content which will reduce drying costs.

Much can be done to maintain output by paying attention to methods of working, so that the combine spends as much time as possible actually cutting a full table width; also to organisation of transport, ensuring that a trailer is always available and the combine does not have to wait with a full tank.

The greatest cost of combining is the depreciation charge. This is very high at low levels of usage; on small acreages, the use of a contractor's services may be cheaper. Against this must be set the possibility of crop losses due to untimely cutting resulting in a greater loss of income than the cost of investing in a second-hand combine or joining a syndicate.

Grain storage

Most grain coming from the combines will be put into some form of storage for sale or use at a later date. The grain handling and storage arrangements are very important parts of the harvesting process, and there must be a relationship between these arrangements and combine capacity. There can be quite considerable variation in costs of storage installations, depending on the type of system and the amount of handling equipment involved. Storage can only be justified if the costs of storing are less than the increase in value of the grain over the storage period. For grain used on the farm, storage costs must be less than the difference between what the grain would realise if sold at harvest and what would have to be paid to buy in grain as and when required.

With the present price structures for wheat and barley, it is possible to increase the gross margin per acre by storage and selling late in the season. Storage costs are likely to be at least £3 per ton, and whilst it is profitable to store wheat, the profit from storing barley can be marginal in some years; therefore costs must be kept to the minimum consistent with efficient operation. In calculating storage costs, allowance must be made for interest on the capital represented by the stored grain. A farmer with a large overdraft may prefer to sell grain to reduce his overdraft rather than store it and pay interest on the borrowed capital. Selling grain direct from the combine makes the farmer entirely dependent on the merchant accepting the grain before it deteriorates, generally at a very low price. It tends to be a rather inflexible system and some storage is normally considered necessary on most farms.

Before grain can be safely stored, it must be treated to prevent deterioration caused by the growth of micro-organisms, i.e. moulds, bacteria and yeasts which are always present to some extent when the grain is harvested. The usual treatment is to reduce the grain moisture content. Other methods include cooling, sealed containers and application of chemicals.

Grain drying

The range of driers available is wide. They can be classified into three basic groups, i.e. continuous flow, batch and storage types; there are a number of variations within each main group.

With the trend towards bulk handling of grain from the combine, driers for grain in sacks are becoming much less important and are now used only on small farms. On larger farms the amount of moisture which normally has to be removed from the grain will have a decisive effect on the type of drier installed. Continuous flow or automatically-controlled bulk batch is usually preferred in the wetter areas, and storage types in drier areas. There is however no clear-cut division regarding this, and storage driers, particularly on-floor types, are becoming increasingly popular. Accurate detailed costings of installation are difficult to give, since there are so many variables: e.g. nature of site; type of building; type, capacity and make of equipment installed; grain handling arrangements. Large installations are not always cheaper per ton than small ones, since they usually require new buildings, whereas small installations can often utilise existing buildings.

Grain storage at high moisture contents

The inclusion of a high proportion of rolled barley in livestock rations has recently created a great interest in the storage of grain at higher moisture contents. For satisfactory rolling, the barley must contain at least 16–18 per cent moisture; and since drying is relatively expensive, it is obviously uneconomic to dry the grain for storage and to wet it again for use if it can be stored safely and cheaply in its undried state.

Storage in "sealed" silos is the most used method at present, and there is a fairly wide range of equipment in use. Most silos are of metal construction, made from either galvanised or vitreous enamelled sheets bolted together and with mastic sealing compound between the joints to make them air-tight. Pressure variations in the silo are relieved by either a valve or a breather-bag inside the silo roof. Unloading is by a sealed auger, which may be simple or supplemented by a sweep auger fitted in the base of the silo.

Some silos are made from butyl rubber supported by a metal cage. This type is cheaper, and since they are flexible they do not require safety valves. They are however more liable to damage, and can present some problems when used out-of-doors, such as an accumulation of rain-water in the hollow made in the top as grain is taken out.

Compared to a grain drying and storage installation, all forms of high moisture content storage silo are relatively cheap. Grain can be put into the silos between 18 and 24 per cent moisture content, and where equipment and management are satisfactory results show that the losses are negligible and feeding value is at least as good as that of grain which has been dried. Development of certain bacteria and yeasts gives the grain a "beery" smell and taste. This makes it unsuitable for milling for human consumption. The germination is destroyed making it useless for malting or seed; however, the grain is palatable and very acceptable to livestock.

As yet there is very little market for this type of grain to be sold off the farm, and this form of storage is essentially one for use where the grain is to be consumed by livestock on the farm.

With this type of storage there is a danger from the accumulation of carbon dioxide gas within the silo. When a worker has to enter a silo containing grain, it is essential that every precaution should be taken to ensure his safety.

Grain chilling The storage of damp grain by cooling it with refrigerated air is in use on a limited scale. Storage may be either in ventilated bins or on the floor. The success of this technique depends upon relating the storage temperature to the grain moisture content. Within limits, grain stored by chilling can be used for seed, malting or milling.

Some refrigerated stores are used in conjunction with batch or continuous driers, either for temporary storage before drying, or for long-term storage after drying to 16–18 per cent. An arrangement such as this can be an asset when the drier capacity is limited, since the capacity of the chilling installation can be modest and the cost of installation would be considerably cheaper than the purchase of a large drier. On 500 tons capacity the cost would be approximately £1 per ton for the installation. Running costs are approximately 5p per ton for the initial chilling plus

2p per ton per month for maintaining the reduced temperature during storage.

Low volume aeration techniques have come into being since about 1969-70. This technique involves sucking or blowing untreated air through the grain in store, so reducing its temperature. Aeration is done mainly at night, taking advantage of the cooler air. Very small ($\frac{1}{2}$ hp) fans are used, and the cost of installation and running is very low (50p to £2 per ton installation cost, depending on tonnage and type of installation; running cost 1p per ton per month). There is a definite upper limit of 18 per cent moisture content if the grain is to be kept in good condition. This type of installation can form a useful addition to an existing drying plant which is of insufficient capacity.

Chemical treatment. During very recent years a technique of preserving grain by treating it with propionic acid has come into use on farms. Propionic acid is related to acetic acid as found in vinegar and occurs naturally in the rumen of cattle. The treatment consists of spraying undiluted acid on to the grain as it passes into the store. Several manufacturers make calibrated auger/spray units for carrying out the treatment. The amount of acid used depends on the grain moisture content. 1 per cent by weight being the dosage on grain up to 25 per cent moisture content; this at present costs approx. £1·50 per ton, but on grain of 18-20 per cent moisture content the cost is approximately half this figure. The grain is given a vinegar-like smell, and is unsuitable for milling; the germination is affected, making it unsuitable for malting or seed.

The merits of the method are that no special buildings are required for storage, the grain being simply piled on the floor. The acid maintains its effect over the whole storage period and after the grain has been rolled in preparation for feeding to livestock. In some areas of Great Britain propionic acid treatment is now the most popular system of storage where the grain is being fed to livestock on the farms.

SUGAR-BEET

Apart from the need to reduce labour costs, farmers in the United Kingdom are faced with the fact that mechanisation of other crops has reduced the supply of regular farm labour available for carrying out hand work on sugar-beet, and methods of partial or total mechanisation are being used increasingly both to cut costs and ensure timely treatments. As machinery replaces hand work, it becomes increasingly important to make sure that each seed has the best possible chance of germinating and emerging to grow into a plant and that the emergence is even. This even braird can only be achieved if attention is paid to detail during the drilling and seed-bed preparation. The foundation of a good seed-bed is laid during autumn ploughing. Reversible ploughs are being used increasingly, since they are the only means of leaving a really level seed-bed. Correct choice of plough mouldboard is important and good ploughing is essential so that the surface is left level and even.

Spring work demands experience and skill to minimise moisture loss and prevent compaction by tractor wheels. Cage wheels on the tractors, and the use of implements in tandem behind high powered tractors, are becoming the accepted practice, and a number of implements designed specifically for sugar-beet seed-bed work have appeared in the last few years.

Almost the entire sugar-beet crop is now drilled by precision seeders. Genetic monogerm seed is used increasingly, but the flat seed shape is such that it does not suit the drills at present in use unless it is pelleted to make the shape spherical. Various forms of drill are available to place the seed at desired spacings and in some cases to desired patterns of spacing. Drilling to a stand has become popular with many growers, but on some soils there are considerable risks of crop failure, and sowing at closer spacing acts as a safety factor. Great care must be used when drilling to put all seed in at an even depth to obtain even emergence. Rows must be kept straight if subsequent inter-row work and harvesting is to be carried out efficiently.

Many farmers use band spraying attachments on the seed-drill. These spray a band of pre-emergent herbicide over the seed row to control weeds which would otherwise grow up in the row competing with the plants and making thinning more difficult. These herbicides unfortunately do not always give 100 per cent control, and this can lead to difficulties where total mechanisation is being practised. Post-emergent sprays which are applied overall are now available and are being increasingly used.

Thinning of the beet to its final spacing is still commonly done by hand; this can take up to 40 man-hours per acre. Mechanical thinners are available in various forms, and although complete machine thinning can eliminate hand work, it can also result in a loss of yield of up to 10 per cent. Some farmers make a compromise, and put the mechanical thinner through the crop once and follow by hand work. This reduces the hand work required by roughly 50 per cent, and allows the hoe men to cover the ground quicker.

A recent development is the electronically controlled selective thinner. This machine attempts to produce a stand which is more regularly spaced than that produced by the original random thinners. There are two basic types—those which remove unwanted plants mechanically, and those which spray the unwanted plants with a contact herbicide. The machine works by means of a sensing device which can be set to a certain distance between plants. If there is no plant at this distance the machine automatically leaves the next one and then reverts to its original setting. A limitation to present machines is that they cannot discriminate between plants and weeds, so the row must be absolutely weed-free.

The chemical spray type has advantages in that the soil is not disturbed and so the pre-emergence spray band is not broken, and the plants are not disturbed and so set back. Both types of machine are expensive but they have an application where drilling to a stand is not possible.

Tractor hoes for inter-row weeding can be either front-, mid- or rear-mounted. The rear-mounted steerage hoe is still preferred by many farmers on account of its greater accuracy and rate of work; it does however require two men. Recently front-mounted hoes have become more popular. Specialised tool-frame type tractors are ideal for this work, but can only be justified on quite large acreages. Recent introductions include tool-carrier frames attached to the front of a standard tractor. These require only one operator who sits on the frame and controls the tractor from this position. This arrangement gives good visibility of the work being done, and enables one man to achieve a high rate of accurate work.

Inter-row sprayers for weed control are used by some farmers.

These consist of low-pressure jets which are caused to vibrate electrically using power from the tractor battery; this results in a coarse spray which will not drift. The plant rows are protected by shield plates.

Harvesting

At least 95 per cent of the beet crop in Great Britain is now harvested mechanically. Hand lifting requires approximately 50 man-hours per acre excluding carting off the field. A two-man system using a harvester can do the entire operation including carting in approximately 7 man-hours per acre. Even on small acreages, a harvester is cheaper than hand lifting at piece-work rates.

The many problems involved in the mechanical harvesting of sugar-beet are mostly created by the working conditions, which can be very wet and sticky. This can cause traction problems; also wet soil building up inside harvesters can cause excessive strains on chains and lead to time-wasting breakdowns. At present, most harvesters are single-row machines which deliver the beet up a side elevator to a tractor and trailer running alongside. This method requires a labour force of at least two men, and there is a steady increase in the use of tanker-type harvesters which can be operated by one man. One-man systems are the most efficient in terms of man-hours per acre harvested, but to obtain the highest output in terms of tons per hour it is usually necessary to operate a two-man system, the harvester discharging its tank into the trailer whilst on the move as opposed to running to a heap or a static trailer to discharge. Self-propelled harvesters are being used increasingly; the British machines utilise the power unit and transmission of a medium horse-power wheeled tractor. This type of harvester is more manoeuvrable than a trailed machine, and can operate at higher speeds.

The disadvantage of mechanical harvesting, particularly on heavy land, is that the harvester and tractors and trailers used for carting off the beet tend to cause rutting and damage to soil structure if used in wet conditions. Multi-row harvesters have been developed, and although these machines are heavier the reduced number of their passes across fields results in less soil damage. The greater output of these machines allows the harvesting operation to be done when conditions are rather more suitable. British-made machines are limited to 3-row operation at present.

Multi-stage harvesting is quite common in Europe, and a number of these systems have been imported into this country during the last three seasons. The machinery lifts up to 6 rows at once and comprises three machines—a topper, a lifter/windrower and a pick-up cleaner/loader. The equipment cost is high and a large labour force is required and the output claimed by the makers is up to 2·5 acres per hour.

Collection of tops for feeding to stock can be provided for on most harvesters by an extra attachment. Some machines collect the tops in a container and discharge at intervals, leaving the tops in a window at right angles to the crop rows; others use a deflector to place 3 or 4 rows into a windrow parallel to the crop rows. Some harvesters can load tops into trailers alongside, but this tends to make harvesting more complicated and requires another man.

Beet taken straight to the factory is liable to show heavy dirt tares if the conditions are wet. The dirt tare can be considerably reduced by clamping the beet on a concrete apron for several days and using a cleaner/loader to load the transport vehicles, the beet being loaded into the cleaner by a tractor-mounted loader equipped with a special root bucket.

POTATOES

In recent years a number of developments have taken place in potato growing techniques. These have been mainly directed at providing clod-free conditions at harvest time. The use of complete harvesters is increasing steadily owing to labour costs and shortages, but present harvesters still have difficulty in dealing with large quantities of clods and as a result their output suffers.

Potatoes are grown on a variety of soil types, and there is no universally sure technique for making and maintaining clod-free conditions. The basic requirements are good ploughing to obtain maximum weathering and cultivation which provides a tilth of 4–6 inches without bringing raw soil to the surface to form clods. The use of cage wheel extensions on tractors whilst doing surface cultivations prevents compaction in the damper soil below. Compaction produces clods during later cultivations. The row width is an important factor in relation to the minimum depth of tilth required. Some farmers are now using 36-inch rows as opposed to 28-inch or 30-inch, so that the soil required to create the ridges can be obtained without deep cultivations between the rows; the increased row width also prevents the tractor wheels from pressing the sides of the ridges and forming clods.

Potato planting

Planting by machine is considerably cheaper than hand work, and also has advantages in that the seed is placed in the soil in better conditions. The main disadvantage of machine planters is the limited output of a single machine compared to the output of a large gang of workers. Average hand planting rate is $\frac{1}{2}$–$\frac{2}{3}$ acre per day per worker of unchitted seed. A good gang with well-organised transport for seed may well achieve $\frac{3}{4}$–$\frac{7}{8}$ acres per day. Rates for chitted seed are a little less.

Planters can be divided into two main groups, i.e. hand-fed or automatic. In the first group each tuber is placed by hand either into a chute or on to some form of conveyor. One operator is required per row, and at normal seed spacings the forward speed has to be limited to approx. 1 mph, owing to the inability of the operators to handle the potatoes any faster. This results in an output of approximately $\frac{1}{4}$ acre per hour from each row on the machine.

Automatic planters pick up seed from the hopper by some form of conveyor, and are independent of human limitations. With suitably graded seed they can work at speeds up to 5 mph which means that a 2-row automatic planter can have an output better than a 4-row hand-fed type. Only the tractor driver is required on some machines, on other machines a second operator is required to supervise the machine and refill the "gap fillers" where these are fitted. Generally automatic planters are considered unsuitable for chitted seed due to the damage incurred by the shoots. However chitting in a controlled environment as opposed to a glasshouse can result in short sturdy shoots which do not knock off, and there are machines on the market which can work well with this type of seed.

Once the seed is in the ground there is considerable difference in techniques used by different farmers. The traditional technique—using cultivation implements to pull down ridges, and then a ridging plough to rebuild the ridges—is still used by many farmers, but there is a great risk of forming clods in the ridge, both by the tractor wheels and by the smearing action of the mouldboards of the ridger. One technique has been developed to ridge up immediately after planting, giving the potato a covering of 6–8 inches of well-worked soil, spraying with herbicide, and doing no further cultivations at all. Another technique goes to the opposite extreme. This is known as the "Dutch" method, and involves the frequent use of a complex implement which scrapes up fine soil from between the rows (usually 36 inches) and throws it lightly on to the ridge, so building up the ridge in several stages as the potatoes are growing. This method requires good management, or it can result in clod formation.

Inter-row rotary cultivators and rotary ridgers have been developed, and although they are expensive they can produce clod-free ridges in one pass and are an asset on certain soils.

Potato harvesting

At the present time, a large casual labour force is employed during the harvesting of main crop potatoes. The labour requirements for hand picking are very variable, depending upon the conditions, the machinery used for lifting and the skill of the pickers. A well-organised, skilled gang, picking after an elevator digger in good conditions, can complete the whole operation of digging, picking, carting and storing in approximately 40 man-hours per acre. But the average gang consists of unskilled workers, and organisation can leave room for improvement, and so the time for the whole operation might well be up to 80 man-hours per acre, costing up to £20 or more per acre in a difficult season.

Two methods are used for picking after spinners and diggers. In the "stint" method, each picker has a length of row. A fairly large gang is required if the digger is to work continuously, and good organisation is essential if output is to to be kept up. In the "breadth" method, an elevator digger fitted with a deflector for offset delivery (or a 2-row digger) is used to lift as many rows as there are pickers. One picker works to each row and baskets are emptied into transport which follows the pickers up the rows. This method can utilise a small gang, and the rate of picking is better than by the stint system due to the pickers always having potatoes in front of them. The method is not suitable for large gangs, and is also more vulnerable to the weather, in that should picking be stopped by heavy rain the fact that the potatoes are lifted ahead of the pickers may result in some not being picked up until the following day.

Picking into baskets, emptying these into bags and loading the bags on to trailers for transport is a laborious and time-wasting method and presents many opportunities for bruising to take place. Emptying baskets direct into trailers is an improvement on this, but can result in very hard work for the man emptying the baskets. Picking into 5 cwt pallet boxes has several advantages. Emptying baskets is easier and handling from field to store can be done without any hand work at all. Pallet boxes can be used in several ways depending on the transport system, i.e. emptied into bulk trailers, or loaded on to flat trailers, or carried on special devices attached to the front or rear of a tractor for transport. A special tipping device attached to a tractor is used to tip the potatoes on to the heap. If this is properly used bruising is kept to a minimum.

Complete harvesters

Casual labour is becoming more difficult to obtain, and the cost continuously rises. It is fairly easy to justify the use of complete harvesters where they can be made to work. Unfortunately the present designs of harvesters do not perform satisfactorily on many of the soils on which potatoes are grown. A machine which will work very efficiently on fen or light loam soils with only two operators on it, may be most unsatisfactory on heavy, wet soils where clods or stones are present, even with six operators. In such conditions the rate of harvesting is very slow and may not be high enough to complete the harvest before the weather reduces conditions to those under which the machine fails completely.

As written earlier, serious attempts are being made to minimise clods in the rows. This, and developments in harvester design, should improve performance. Several methods of automatic separation of potatoes from stones and clods are now in use, the latest development being the use of X-rays and electronics. Under ideal conditions the labour requirement for the complete operation of lifting, carting and storage can be as low as 20 man-hours per acre, but a more likely figure is 40 man-hours per acre. Even so this compares very favourably with up to the 80 man-hours per acre required by the average hand-picking gang. The average performance rate of a single-row complete harvester will be $1\frac{1}{2}$ to 2 acres per day; a two-row harvester will nearly double this output.

A 2-stage harvesting system has been developed by one manufacturer which makes use of modified 2-row elevator diggers. The harvester picks up 6 rows at once by means of a special lifting share. Further cleaning is usually necessary before the potatoes go into store. The system can operate at very high rates, e.g. 15–20 acres per day, with a labour requirement of 6 man-hours per acre, but it is only suitable for stone-free soils.

CONSERVATION OF GREEN FODDER CROPS

Haymaking is by far the most widely used method of conservation at present, accounting for some 86 per cent of the total dry-matter conserved. Silage-making accounts for approximately 13 per cent, and grass drying for approximately 1 per cent.

Haymaking

Many important developments in both machines and techniques used for haymaking have taken place. These are almost all concerned with speeding up the curing process, since shortening the period of exposure to the weather invariably results in an improvement in the quality of the hay.

The principle of severe treatment of the swath at or soon after the time of cutting is becoming generally accepted. Although initially crimpers and roller crushers were developed for use after reciprocating knife mowers, the more recent development of flail mowers shows greater promise. The cost of a flail mower is approximately the same as a reciprocating mower and conditioner, the resultant swath treatment is similar and the output is likely to be better, particularly in heavy crops. The disadvantage of flail mowers is their high power requirement, a 5 ft cut machine requiring at least 50 hp in thick crops.

Rotary drum mowers also work well in difficult conditions. The treatment of the swath is not severe, but the method of cutting leaves the swaths in an ideal state for continuous tedding. The technique of continuous tedding, 4 or 5 times a day in good weather, can result in quick drying; but care is necessary when the crop is nearly dry, or a severe loss of leaf is possible.

Most hay at present is handled by pick-up balers. Although the handling of bales presents some difficulties, efforts at handling hay loose in a chopped state and at wafering have so far met with little success. The most common form of baler is the twine-tying ram-type making bales of 18 × 14 inches cross-section with a length variable from 24–42 inches.

Bale handling

Many methods of handling bales from field to store are in use. Handling bales singly has a high labour requirement and can be heavy work. It is also necessary to keep close up behind the baler in case of rain which can cause considerable damage. Unless the haulage team can keep up with the baler, the baling operation may have to be slowed down, and for the same reasons the use of manned sledges towed by balers has declined. Modern balers, given suitable conditions, can turn out 8 or 9 bales per minute, and as no man can keep up with this rate of work, the operation has to be reduced to a speed more in keeping with the capabilities of the man on the sledge. Unmanned windrowers followed by hand-stacking into groups for mechanical handling have become very popular, but automatic grouping sledges leaving bales ready for mechanical handling are being used increasingly due to their lower labour requirements. Methods of moving bales from the field depend very much on transport distances; over short distances tractor-mounted carriers are very efficient; over greater distances or on public roads trailers must be used, the best type being fairly large and with retaining rails at front, rear and one side.

Hay drying

Artificial drying of hay not only removes much of the risk from the haymaking process, but it can also result in a higher yield of a better product. Loss of crop due to leaf shattering can be quite high in field-finished hay. If the crop is baled at a higher moisture content, shattering is much less, and since the leaves contain the highest proportion of protein the quality is improved. The increase in yield of material might be worth up to £3 per acre, apart from any increase in quality. The actual cost of drying depends on the buildings, type of installation and equipment used, and the amount of moisture being removed. Running costs will vary from 50p to £1.00 per ton; fixed costs must be kept below £2 per ton if the operation is to be economic.

Hay drying can be done on a batch or a storage basis, the method used depending largely on the buildings available and the amount of moisture which can be removed in the field. The batch method entails double handling, but is more effective on high moisture content material. A range of fans is available using electric motors, diesel engines or tractors to drive them. Some driers are fitted with heaters which may be electric or oil-fired indirect heat exchangers, or may make use of waste heat from the engine driving the fan. Drying costs can be kept down by using a large volume of cold air, but protracted drying can cause losses of dry matter and the addition of a little heat can speed up the drying process. The waste heat from either a diesel engine or tractor engine driving the fan can raise the temperature of the drying air by up to 4°C (8°F), depending on the horse-power developed by the engine and the air flow of the fan. This is only a small rise in air temperature, but it is sufficient to ensure satisfactory drying in most conditions. Using electrical heaters temperature rises of up to 11°C (20°F) are possible, but the costs are considerably greater.

A good standard of field work is essential. The moisture content must be reduced to 45 per cent or less in as short a time as possible, and correct and uniform density of the bales is also important to ensure even and economical drying.

Grass drying

Something like 100,000 tons of dried grass are produced annually in Great Britain, most of which is sold off the farm for use in poultry foods where its carotene content is of great importance. The use of dried grass for cattle food is more difficult to justify in view of the fact that results which are nearly as good can be obtained much more cheaply by hay drying. Recently, however, owing to the higher costs of protein concentrated foods interest has been revived in grass drying for cattle food, and machines have been developed to make the grass into wafers without prior grinding (which was the accepted method previously).

Modern driers are expensive both to buy and run and can only be economic if fully used. Farm-type driers which are also used for drying grain have a low throughput, e.g. 3–5 cwt per hour. Such an installation would cost approximately £10,000 for the drier and related equipment. A high temperature pneumatic drier of 2 tons per hour throughput together with processing, field and transport equipment might cost nearer £70,000. This type of drier is much more efficient, but obviously requires to be part of a large-scale enterprise where it is kept in use for a considerable part of the year before it becomes economic.

Silage-making

The introduction of the flail-type forage harvester to Great Britain had a very great effect on silage-making. Using modern harvesters and powerful tractors, high rates of work are possible, and this reduces the labour required and can produce a better final product.

A high rate of working is essential, so that the crop can be ensiled whilst at its best, and whilst the weather is fine, and the silo filled quickly. Three types of harvester are available at present: the simple flail, the double-chop, and the metered or precision-chop. The simple flail and double-chop types are used for clamp and pit silos; the double-chop is more expensive but provides a shorter-cut material, which makes better use of trailer capacity and is easier to handle at the silo. The precision-chop machines are intended for use with tower silos where a short chop is essential for mechanical unloading, but a few farmers are now using this type of machine in conjunction with walled clamp silos (bunkers) where mechanised unloading is practised as opposed to self-feeding.

The choice of system depends on the number of men available, and the rate of working and type of silage required, i.e. direct-cut or wilted. Wilted grass usually makes better silage but requires an extra operation and more labour, which may not always be practicable especially on smaller farms. Systems are in use involving from one to five men. Generally large gangs are less efficient than small ones, and a three-man gang is possibly the optimum for direct cutting with a four-man gang the optimum on wilted material.

High dry-matter silage. Tower silos are used by some farmers. The principal advantages are: the good exclusion of air from the crop during ensiling and storage, so that fermentation and losses may be controlled, resulting in a greater efficiency of conservation; and mechanisation of the filling and emptying operations, resulting in some saving of labour and saving of waste when feeding. The disadvantages are: extra cost for machinery both in the field and at the silo; the unreliability of some machines; and the difficulty of wilting down to the required level of 40 per cent dry matter in some seasons. Owing to the machinery costs, the tower silo method only becomes economic on fairly large installations, for instance for a herd of 80 cows and over. Correct planning of the installation and good management are essential if the system is to reach its potential in improvement of conservation, saving of waste and saving of labour.

Two forms of silo are in general use: concrete stave (unsealed) and vitreous enamelled steel plate (sealed). The vitreous enamelled silos are usually more expensive, but the higher costs should be offset by greater efficiency of conservation due to sealing. Towers can also be divided into top and bottom unloading types. Most silos in the United Kingdom are of the top-unloading type. The reasons for this are that bottom unloading is generally more expensive in terms of capital cost, and that it is essential that the silage must be of the correct dry-matter content for efficient unloading. This can cause wilting difficulties and field losses. Bottom unloaders have certain advantages, such as lending themselves more easily to gas-tight sealing of the tower, and allowing all servicing on the unloader to be carried out at ground level.

MECHANICAL FEEDING

Several forms of mechanical feeder are now available for silage, fodder and concetrates, and there is a definite trend towards the types using endless chains and belts. The advantages of these types are: less wear than on auger and tube types; and the ability to feed mixtures along their full length without causing separation. The disposition of the feeder relative to the silo is a very important point, since primary linking conveyers can be expensive in cost and maintenance. The possibility of future expansion must also be taken into consideration.

Some farmers utilise a mobile forage box drawn by a tractor to fill the mangers. This method keeps down capital cost and makes fuller use of existing equipment, but it requires more labour than a fixed installation, and this can be of some importance. It also requires more space in which to manoeuvre, and if feeding is done inside this can be more expensive to provide than a fixed form of feeder.

THE FUTURE

Following the present trend, there is little doubt that the power of tractors is likely to increase. Hydrostatic transmissions allowing stepless speed variations are becoming available. Work is being done at present on automatic guidance systems for tractors, and there would seem to be a real possibility of this becoming a commercial proposition within the next decade. The use of driverless tractors on routine and transport operations has obvious advantages. The perfect accuracy with which a tractor equipped with a guidance system follows its mark could also be a very considerable advantage in operations such as spraying and fertiliser distribution. Other applications of automatic control of field machinery are also possible. A grain loss indicator for combines is now available. There is a possibility that some combine controls could be linked to such a device, and this would lead to greater efficiency and a less fatiguing job for the operator.

Around farm buildings, automation is already being applied to many operations. Completely automated feed preparation plants are now in use on both large and small farms, and an extension of their use seems likely, since they can reduce or eliminate the labour required for a messy and often heavy job. Automatic feeding systems for all types of stock are coming on to the market. Milking parlour feeders which ration according to yield of milk have certain advantages. For those people who wish to record yields, etc., the recognition of cows by electronic means is being studied; if it is considered necessary to keep full histories of each animal, this recognition system could be linked to a simple computer so that all information would be available when required. Devices for indicating that milk flow has ceased have been in use for some time; a logical development from this is to remove the teat cups by automatic means; development work is proceeding along these lines.

The automatic feeding of pigs and automatic control of ventilation of pig houses is now widely used. Automatic cleaning systems are available and are likely to be used more in the future. The basis of all this "farmstead automation" should be to allow the stockman more time to exercise "stockmanship", rather than spending his time on laborious, unskilled chores.

Materials handling is a subject likely to receive more attention in the future. Already handling in bulk is replacing handling in sacks for goods like grain and feeding stuffs. The use of pallets, and handling by fork lift trucks or tractor attachments, is becoming more widespread for products leaving the farm, such as potatoes and vegetables.

All equipment is expensive to buy and maintain at the present time, but without it labour costs would inevitably be exceedingly high. The economic aspects of mechanisation and of the other systems of crop and stock production dealt with in this book are discussed in the concluding chapter.

REFERENCES

1. Helme, W. H.: *Labour and machining costs in the S.E. region.* NAAS (1970)
2. Sturrock, F. G.: *Report on farming No. 46 (1957)*, Cambridge University School of Agriculture
3. Nix, J. S.: *Farm management pocket book*, Wye College Dept. of Agriculture Economy

Chapter 12

ECONOMICS OF BRITISH FARMING SYSTEMS

The preceding chapters have described the husbandry aspects of the farming systems commonly found in Great Britain. The importance of following the rules of good husbandry cannot be over-stressed. However the choice of system will ultimately depend on the personal preference of the farmer, and the economics of the possible enterprises. The purpose of this concluding chapter is to introduce the reader to the economics of the various farming systems as a basis for the exercise of choice.

DEFINITIONS AND CLASSIFICATION

To provide a meaningful discussion on the economics of these different systems it is necessary to define clearly the terminology to be used. Due to the rapid development of farm business management techniques over the last ten years, there exists a lack of uniformity in the definition of terms used. The gross margin system is the most popular, but as many farmers are only now changing to a gross margin form of accounting little published data are available nationally on which to base a comparison of the different farming systems. Thus in this chapter it has been decided to use the Gross Output/Cost method of presentation.

The definitions used are as follows.

Gross output represents the Sales plus any direct grants or subsidies and after allowing for any valuation changes less any purchases of livestock, livestock products or any other products bought for eventual resale. The Gross Output therefore represents the volume of production from the farm business during one year's trading. One way of increasing production is by purchasing feedingstuffs and seeds needed for livestock or crop production rather than growing these resources at home. In order to obtain a better measure of the production on a particular farm it is more informative if the cost of purchased feeds and seeds is deducted from the Gross Output. The resulting term is known as *Net Output*.

When the running costs and overhead costs are deducted from the Gross Output the balance is the profit or loss for the year. The term *Profit* has so many meanings that it is now standard practice to use the term *Net Farm Income*, particularly as this is much more explicit than profit.

To compare the financial results of different systems it is obvious that these results must first be prepared on a similar accountancy basis. Two main problems are involved here. Firstly, does the farmer own or rent the farm; and secondly, how much manual work does he do on the farm? To standardise the accounts for comparative purposes all farms can be treated as tenanted farms. For owner-occupiers this means the addition of a realistic rent figure and the deduction of those expenses normally borne by a landlord, e.g. drainage rates, building repairs, tithe, and insurance of buildings.

As the Net Farm Income represents the farmer's reward for his manual work, the investment of his capital, and his management of the holding, the extent of manual labour he puts in will affect the size of the Net Farm Income. To make the farms comparable the value of the farmer's manual labour and that of his wife is usually deducted from the Net Farm Income to produce the income he receives for his management and investment.

Thus:

Gross Output — Costs = Net Farm Income
Net Farm Income — Manual Labour of Farmer and Wife = Management and Investment Income

To assess the return that the farmer (as a tenant) is obtaining from his farming business it is usual to represent the *Management and Investment Income* as a percentage of the *Capital* that he has invested as a tenant (including any borrowed capital). This figure is known as the *Return on Tenant's Capital*. The definition of Tenant's Capital is, however, perhaps the most controversial topic in present-day farm management.

For this discussion the average valuation of crops, produce, and livestock plus the book value of machinery and equipment is used together with an arbitrary allowance for working capital, i.e. the capital needed for the short-term financing of the farming operations such as purchase of seeds, feeds and fertilisers up to the time of sale of the crops or stock. Obviously the date of the end of the farm's financial year will have an important influence on the valuation. Most farmers undervalue by about 10 per cent on livestock, and 5 per cent on total valuation on the premise that it is unwise to anticipate profits by using figures which may not be achieved when the crop or animal is eventually sold.

The type of system and combination of enterprises will also affect the amount of working capital required. In theory an enterprise with a regular sales pattern, e.g. milk production, should require less working capital than, for example, 18-month beef in which two batches of animals will have been purchased and fed before the first cheque is received eighteen months after starting the enterprise. In practice, however, the use of mixed farming systems, of merchants' credit, or of crop discounting tends to even out the flow of cash. On the average farm the working capital would be between 8 and 15 per cent of the annual valuation, assuming that the latter is realistically prepared.

When examining the *Farmer's Income* it is important to remember that the farmer could earn a Gross Return of 8–10 per cent per annum by investing his capital in a building society or similar savings scheme. This annual income could be obtained with the minimum of work and risk. If, therefore, he decides to invest his capital in farming he will require a much higher gross return to repay him for his investment, his manual labour, and his managerial ability. The income that he receives for the manual and managerial functions can be calculated by deducting an interest charge of 8 per cent, on his total investment including borrowed money, from the Net Farm Income.

Thus:

Farmer's Income=Net Farm Income less Interest at 8 per cent on Capital Invested.

The interest charge represents either the net interest he would be charged by his bank if he was farming on borrowed capital after allowing tax rebate on this interest charge, or the Gross Return he would get by investing his capital in a building society.

The *Farmer's Income* is his reward for his manual labour and his management; it is available for providing for taxation, living expenses and for reinvestment in the business. The present cost of living, the present taxation system, and the rapidly increasing cost of replacement of assets mean that even the smallest farm business will require a minimum Farmer's Income of approximately £2,000–£2,500 per annum if it is to survive for any length of time. An even greater sum will be required if the farmer has also to repay any borrowed capital such as a mortgage, a long term loan, or his bank overdraft.

In comparing farming systems and farms within a particular group, some classification according to farm size is usually required. With the increasing scarcity of skilled labour, and with the farm labour force expected to decline by one-third in the next 20 years, it is clear that labour may become more important than acreage in limiting the expansion and profitability of the different farming systems. Therefore the classification according to labour requirement may be more valuable than one based on acreage, particularly when the very wide variation in the productive capacity of land in different regions is taken into account.

The usual method of assessing the labour requirements of a farm or farm type is to use *Standard Man Days** i.e. the number of days' work by an average male employee required to produce one acre of crop or care for one unit of livestock, assuming that he works an 8-hour day. Obviously the requirements will be conditioned by degree of mechanisation, topography, farm layout, buildings, skill of workers and efficiency of labour organisation. However each region has a list of Standard Labour Requirements published regularly by the Regional Agricultural Economics Department of the local University, or in Scotland by the Colleges of Agriculture. Under average conditions one man would provide 275–320 man days per annum—the higher figures usually being associated with stockmen and dairymen who work longer hours than arable workers.

The classification used here is:

275-599 man days. Farm employing one or two men full-time plus part-time assistance. One of the men on the smaller farm is likely to be the farmer himself.

600-1199 man days. Two to four men. On arable farms, the farmer would be doing less manual work and would be more involved in the managerial aspects. Some part-time and seasonal help.

1200-1799 man days. Four to six men. Possibly much part-time or seasonal work.

1800 man days and over. Over six men, with the farmer only working at peak periods.

* Often abbreviated as SMD.

The classification is also based on farm type. Figures for nine systems are given.

(1) *Specialist dairy farms*. These are farms whose main enterprise is milk production, all other enterprises requiring less than 25 per cent of total standard man day requirements; usually all-grass farms with possibly some feed roots grown (kale, turnips, etc.).

(2) *Dairy with cereals*. Farms on which Dairying requires between 50 and 75 per cent of the standard man days. Cereals generally form the other main enterprise.

(3) *Sheep farms*. Sheep require more than 75 per cent of labour requirement. Generally these farms would be in the poorer, marginal districts. Very few lowland farms are devoted so exclusively to sheep production.

(4) *Beef and sheep farms*. Farms on which livestock rearing requires more than 50 per cent of total labour requirement. This includes some all-grass farms and some farms which grow a small acreage of cereals.

(5) *Extensive cereal farms*. More than 50 per cent of the farm in cereals with some cash cropping.

(6) *Cash cropping*. Farms growing potatoes, sugar-beet or vegetables as major enterprises, together with high proportion of cereals. Total cropping more than 50 per cent of labour requirements.

(7) *Mixed farming*. More traditional system with no one enterprise requiring more than 50 per cent of labour requirements.

(8) *Pig and poultry holdings*. The small units would be largely accommodation land for the buildings. The larger units may also include some cereal production with the grain being fed to the livestock.

(9) *Horticultural holdings*. These are discussed in more detail in Chapter 8.

Source of information

All figures are taken from or based on the Ministry of Agriculture,

ECONOMICS OF BRITISH FARMING SYSTEMS 205

Fisheries and Food publication *Farm Incomes in England — Wales, 1968*. They refer to the financial results of over 2000 farms for the years 1966–1968. At the time of writing these figures were the most recent national statistics available. An asterisk has been used to denote figures for which no information is available.

SPECIALIST DAIRY FARMS

With the prevailing price structure of the various agricultural commodities, milk stands out as the one commodity that can produce an income in excess of £50 per acre every year. Most other commodities either fall short of this, or cannot be produced continually on the same land without a resulting drop in physical and financial yield. Hence it is not surprising that in a country where the average farm size is only a little over 100 acres, milk production plays such an important part in the agricultural economy. In certain counties such as Somerset, Cheshire, Devon and Dorset, the majority of farms would be less than 150 acres. If the farmer, whether an owner-occupier or tenant, is to make a living even equal to that of manual workers in industry, it is clear that his main enterprise must be capable of producing a regular substantial cash income to cover his overheads, to provide for taxation and leave him enough to live on.

In parts of the North of England the acreage of land is not as limiting as its productive capacity. In many of the fell and dale areas of Yorkshire, Durham, Westmorland and Cumberland many farmers are forced into milk production because the land is not suitable for cropping and the prevailing prices for other forms of grazing livestock production are too low to generate sufficient income to leave a substantial profit unless the farm is over 250–300 acres.

The specialist milk producer is often wholly committed to that form of production. From Table 12.1 it can be seen that three-quarters of such farms are under 200 acres and that the average acreage of all the farms in the survey was only 103 acres. One-third of these farms are probably only employing part-time help—the farmer and his wife providing most of the labour required. Even on the larger units (over 1800 SMD) the farmer is working almost full-time on manual tasks. The following tables will show that only beef and sheep farmers put in as much physical effort as the large specialist dairy producer.

Table 12.1
Specialist dairy farms

Item	Units	Business size (standard man days)					Average of all farms
		275–599	600–1199	1200–1799	1800–2399	2400–4199	
Average acreage	Acres	67	130	213	330	*	103
Average labour required	SMD	439	834	1382	2020	*	665
Distribution of holdings	%	31	46	16	5	2	—
Gross output per acre	£	70·1	71·1	74·5	71·6	*	71·0
Net output per acre	£	46·0	48·4	52·5	52·0	*	48·0
Net Farm Income per farm	£	1130	2037	3375	4502	*	1647
Net Farm Income per acre	£	16·9	15·7	15·8	18·7	*	16·0
Manual labour of farmer and wife	£	714	714	590	561	*	702
Management and Investment Income per farm	£	416	1323	2785	3941	*	945
Management and Investment Income per acre	£	6·2	10·2	13·1	11·9	*	9·2
Tenant's capital per acre	£	68·0	68·4	73·7	74·3	*	67·2
Return on tenant's capital	%	9·8	14·9	17·8	16·1	*	13·7
Farmer's Income	£	793	1238	2120	2541	*	1094

* These figures are not available

However the picture is by no means black. As mentioned above the Gross Output from milk production is generally above that from any other system with the possible exception of intensive potato and sugar-beet growing. With Gross Output and Net Output figures of £71 and £48, the farmer is in a fairly good position to be able to pay all the running costs of the business and still produce a reasonable Net Farm Income. This is borne out by the average Net Farm Income figure of £16 per acre—a level rarely attained by other land-using systems except some cash cropping farms.

Although this Net Farm Income figure appears quite satisfactory, three factors have to be considered. Firstly, if the farmer is to have a reasonable standard of living his Total Net Farm Income is more important than the per-acre figure. The Net Farm Income has to cover tax payments and reinvestment as well as personal living expenses. In many cases interest charges on borrowed money will also be involved. From the survey three-quarters of the farms were not achieving the target of a minimum Farmer's Income of £2,000 per annum. This would suggest that they could earn more by working in industry and also that, on the larger farms, they were paying their cowmen more than they were paying themselves. However, the motives behind any person's decision to farm, be it dairying or any other form of production, are more numerous than just the desire for a high cash income. Personal satisfaction, perquisites (which may be worth at least £500 per annum for house, car, milk, etc.), capital appreciation, family tradition, opportunity, reduction in death duties and various other factors are involved.

Secondly, in order to produce a high Net Farm Income from a limited acreage, the capital investment in extra livestock and in buildings to house these animals over the winter months is usually much higher than in other farming systems. As the comparison is based on "tenant farmer" figures the Tenant's Capital figures in Table 12.1 do not include buildings. With building costs rapidly rising, it is increasingly more difficult to erect a new dairy unit for less than £100 per cow if all services are to be provided. Thus the average tenant's capital investment at £60–75 per acre could be doubled on an owner-occupied holding if new buildings were required. For a new entrant into dairying under present conditions the cost would be as shown in the following table.

	£	
Land	250 per acre	
Buildings	66 ,, ,,	(£100 per cow on 1½ acres)
Livestock	100 ,, ,,	
Machinery and Equipment	30 ,, ,,	
	446	

The total of £450 per acre, at an interest rate of 8 per cent, would require a cash income of £36 per acre to pay the interest charges before considering capital repayment. Thus it is obvious that the specialist dairy system is a very capital-intensive system, and the high levels of Net Farm Income and Return on Tenant's Capital are essential if the farmer is to be able to increase the net worth of his business.

Finally due consideration must be given to the social aspects of this type of system. With the farmer providing much of the labour himself and with the 7-day week still necessary for dairy cows, the standard of living and level of income must be sufficient to compensate for the lack of leisure time. On many of the smaller units, and perhaps on over one-half of all the farms in this group, relief milkers would not be available to permit the working of even a 6-day week. The recent growth of the Farmer's Relief Services providing temporary replacement staff has helped this, but the majority of services only provide relief on a weekly or fortnightly basis. This does at least allow the farmer to take an annual holiday.

DAIRY AND CEREAL FARMS

With the high capital requirements of an all-dairy policy and the reasonably stable and profitable results from cereal production, many farms have developed a balanced dairying and cereals policy. Under prevailing price levels a system incorporating wheat and barley production coupled with a dairy herd on the rotational and permanent grassland is generally accepted to offer the best opportunity for both a high Net Farm Income and high return on tenant's capital. With concentrate feedingstuffs forming up to one-third of the cost of milk production, the growing part of the concentrates on the farm together with the supply of bedding material as a by-product obviously offers a balanced system. Self-sufficiency however is not necessarily always profitable, as in many cases the purchase of feedingstuffs is one way of increasing the effective size of the business. In terms of cost of production it is generally cheaper to produce on your own farm than to purchase, but the opportunity of producing a more valuable commodity on your farm may well outweigh the cost of saving in growing rather than purchasing feedingstuffs. The feeding of straw and a urea-based supplement in place of hay or silage for the maintenance ration, or possibly concentrates for part of the production ration, is a more expensive method of feeding the cows, but the income from the land released from fodder production, or the possibility of keeping more cows on the same acreage, may result in a higher Net Farm Income than a self-sufficient policy.

From Table 12.2, the first point of interest is that the average farm acreage within each group is higher than in the specialist dairy group (Table 12.1) and the average acreage for the whole group is 159 compared with 103. This is a fairly logical increase as the cereal crop obviously needs less labour than the dairy enterprise, and one man can look after a larger acreage. Over 80 per cent of the farms are over 100 acres and would be employing two or more men on a full-time basis.

With a smaller percentage of the land devoted to the dairy unit the Gross Output and Net Output per acre are below those for the specialist dairy producers. The level of Net Farm Income for each business size is generally comparable but the per-acre figures are lower due to the larger acreages.

In this group the farmer is still working full-time on the smaller holdings but is able to devote more time to his managerial duties as the size of business increases. The farms of less than 100 acres can only be regarded as marginal in that the Management and Investment Income, Return on Capital and Farmer's Income are so low as to suggest that if cash is the main consideration these producers would be financially advised to give up farming.

With the increased acreage of cropping the capital investment is lower then for the specialist dairy farms but still averages over £55 per acre to produce a return of 11–18 per cent on the majority of farms.

Table 12.2
Dairy and cereal farms

Item	Units	Business size (Standard man days)					Average of all farms
		275–599	600–1199	1200–1799	1800–2399	2400–4199	
Average acreage	Acres	93	152	269	393	593	159
Average labour required	SMD	478	849	1427	2004	2892	847
Distribution of holdings	%	21	38	21	11	9	—
Gross output per acre	£	51·6	63·9	61·5	59·3	56·1	59·3
Net output per acre	£	36·1	43·5	45·3	46·9	43·4	42·3
Net Farm Income per farm	£	1291	2156	3012	4872	5778	2064
Net Farm Income per acre	£	13·9	14·2	11·2	12·4	9·8	13·0
Manual labour of farmer and wife	£	703	692	551	453	494	667
Management and Investment Income per farm	£	588	1464	2461	4419	5284	1397
Management and Investment Income per acre	£	6·3	9·6	9·1	11·2	8·9	8·8
Tenant's capital per acre	£	53·4	62·1	64·7	62·0	56·2	59·9
Return on tenant's capital	%	11·8	15·5	14·1	18·1	15·8	14·7
Farmer's Income	£	894	1402	1595	2932	3044	1298

SHEEP FARMS

Due to the technical problems associated with very high stocking rates, it is uncommon to find farms on which the Gross Output from a sheep enterprise is sufficiently competitive with other enterprises to make it the main enterprise on good land. Of all farm animals the sheep is the one with which farmers and scientists have been less successful in developing adequate methods of control of parasites, viruses and other organisms causing reduction in performance of an intensively managed large sheep flock. Consequently lowland sheep are either found as a supplementary enterprise on arable farms or combined with beef or dairy cattle on grass farms. The Romney Marsh is probably the only area where lowland sheep can be found as the main enterprise. The major proportion of the National sheep flock is found on the marginal upland farms or on the hills, fells and dales of Wales, Scotland and northern England.

Mixed stocking with cattle will be discussed in the next section; so the following remarks will largely refer to upland and hill sheep systems.

From Table 12.3 a very sorry picture emerges of the average sheep farm. The Gross and Net Output per acre are extremely low, the farmer works long hours even on the larger farms, and the Net Farm Income is barely sufficient to cover living expenses on 80 per cent of the holdings.

The main factors affecting the profitability of sheep production are:

(a) Stocking rate,
(b) Lambing percentage,
(c) Lamb sale price.

Table 12.4 indicates the wide variation that can occur in these three factors and also in the total income from the sale of lambs.

This variation is quite dramatic. The majority of the lowland farms would have lamb sales of £30–£40 per acre, and the hill farms £3–£5 per acre. The two extremes are not so widely found. The very intensive sheep systems are generally associated with an arable/rotational grassland policy, where the available acres are intensively used. At the other extreme, sheep flocks are extensively grazed on the very poor hill land found in parts of the west and north of Scotland, Wales and the north of England. In these sub-marginal areas, lambing percentages can be very low, and in a bad year, the number of ewe

Table 12.3
Sheep farms—mainly on poor land

Item	Units	Business size (standard man days)					Average of all farms
		275–599	600–1199	1200–1799	1800–2399	2400–4199	
Average acreage	Acres	274	557	988	*	*	462
Average labour required	SMD	470	819	1460	*	*	702
Distribution of holdings	%	29	50	12	9	*	—
Gross output per acre	£	10.7	8.9	9.7	*	*	9.2
Net output per acre	£	8.0	6.9	7.7	*	*	7.1
Net Farm Income per farm	£	769	1265	2814	*	*	1144
Net Farm Income per acre	£	2.8	2.3	3.0	*	*	2.5
Manual labour of farmer and wife	£	717	656	664	*	*	688
Management and Investment Income per farm	£	52	609	2150	*	*	456
Management and Investment Income per acre	£	0.2	1.1	2.3	*	*	1.0
Tenant's capital per acre	£	13.7	13.4	14.7	*	*	13.2
Return on Tenant's capital	%	1.5	8.2	15.7	*	*	7.6
Farmer's Income	£	460	659	1714	*	*	648

* These figures are not available

lambs produced may be less than required to maintain the flock in regular ages. This, coupled with low prices for lambs sold over a number of years and rising costs, has led to a decline in the sheep population, and a considerable area of land on hills and upland marginal areas has been sold to the Forestry Commission or to private Forestry Investment Groups for tree planting.

In these districts the major source of income is the Hill Sheep Subsidy and many farmers would not be able to survive if it were not for this, and also the Hill Cow Subsidy. To this extent it can be argued that these two subsidies are largely a social subsidy rather than an incentive to increased productivity, particularly as the subsidies are paid on the number of stock carried and not on the number of progeny sold. The increase in productivity in many poorer areas has been relatively insignificant except for one or two isolated examples, whereas the relative subsidy payments have been running at £20 million per year.

The traditional concept of the hill farmer producing store lambs for the arable farmer to fatten, and the sale of draft ewes to the lowland flocks for cross-breeding may become outmoded. The low profitability of sheep on arable and lowground farms and the subsequent drop in demand for stock from the hills is forcing the hill farmer to consider fattening his lambs on the farm or to co-operate with a lowland farmer on some formal profit-sharing basis.

The two main problems associated with the store lamb trade are the wide variations in liveweights at auction, and the tendency for footrot and other foot troubles in the lambs after a short period on arable land. The former has been countered to a certain extent by selling the store lambs over a weighbridge at the auction market or privately through a marketing group. The latter, however, still results in a lot of hard work for the arable farmer at a period of the year when he is already working to capacity with the cereal and root harvests and autumn drilling.

With regard to stocking rate, the main limiting factor is the acreage of land capable of growing crops for conservation as winter forage. As lambing percentage and lamb growth rates are strongly correlated to proper nutrition of the ewe, much attention is being paid to methods of land reclamation and pasture improvement. In many areas the ewe would not be physically capable of rearing more than one lamb successfully, so a lambing percentage of 100–120 per cent is all that is required; this can be achieved by ewe selection under a properly controlled breeding programme, with proper utilisation of pastures (and purchased feeding if required) to maintain adequate nutrition.

Table 12.4
Measures of performance of different sheep systems

	Sheep systems on			
	Intensive lowground farm	Good upland farm	Good hill farm	Poor hill farm
Stocking rate—ewes per acre	6	3	1	0·33 or less
Lambing percentage	180	150	100	75
Lambs per acre	10·8	4·5	1·0	0·25
Lamb sale price	£7	£6	£5	£4
Total lamb sales per acre	£75·60	£27	£5	£1

From Table 12.4 it is clear that the hill sheep producer is in a very poor position financially when compared with the other groups. The Output is extremely low and on the smaller farms the Net Farm Income is insufficient to cover even the manual labour of the farmer. As the tenant's capital requirement is low due to the lack of any expensive machinery, the return on Tenant's Capital in some cases is quite high; but even a 100 per cent return would not be satisfactory if the level of income was below that needed for a decent standard of living. From the table it can be seen that 80 per cent of the farmers were left with less than £1000 per year after meeting the capital charges. Hence the commonly accepted view that hill sheep farming is a vocation—particularly as the standard of living and social life of the farmer's family is so far below that demanded by workers in other industries or localities.

BEEF AND SHEEP FARMS

On the better hill land, the uplands and the poorer arable land, mixed stocking with sheep and cattle is still very widely practised.

The hill and upland cattle population consists mainly of suckler cows to produce and rear one suckled calf per annum. The calves generally are sold at 4–6 cwt in the autumn to arable farmers for fattening over the winter period in yards or courts on arable by-products. With subsidies amounting to over £30 per cow and calf, and the calf selling for £10–£12 per cwt, the Gross Output per cow on the better managed herds can almost reach that of the poorer dairy herds, e.g. £100 per cow. Hence there has been considerable increases in suckler cow numbers and in herd size, particularly in parts of Scotland. With modern layouts, and equipment, herds of 200–400 cows looked after by two or three men and fed on silage as their basic winter feed, are possible. As with hill sheep systems, pasture improvement and better cow nutrition are the main factors under consideration at the present time along with the excellent work on performance testing of bulls carried out by the Beef Recording Association (now part of the Meat & Livestock Commission). On the majority of farms where a suckler herd is carried, a sheep flock is also kept, and the mixed stocking of the two is a good aid to pasture improvement. Farm planning on such farms is generally based on what one man can handle, and the common standards now used on the more productive farms are:

1 man— 30 to 50 cows and 400 to 600 ewes
2 men— 60 to 100 cows and 600 to 800 ewes
3 men—100 to 200 cows and 800 to 1200 ewes

The main problem with the intensification of the suckler herd, particularly on lowland, is that of capital investment. The cow and calf plus an allowance for a part share of the bull may involve a capital investment of £100–£150 per cow depending on breed. When related to stocking rate this is even more illuminating (see Table 12.5).

To this capital investment must be added the other items of Tenant's Capital, e.g. Machinery, Equipment, Crops and Stores, etc., and if interest is charged at 8 per cent an overhead cost of £2·64–£12·00 per acre must be covered by the beef enterprise.

On the arable farm a stocking rate of one acre per cow is often suggested as the target for efficient technical production, but the

Table 12.5
Capital investment per acre for various stocking rates

Stocking rate	Capital Investment per acre	
	Cow, calf and bull at £100	Cow, calf and bull at £150
3 acres per cow	33	50
2 ,, ,, ,,	50	75
1 ,, ,, ,,	100	150

financial implications, i.e. a capital charge of up to £12 per acre required to service the capital involved, are often forgotten.

The high capital requirement coupled with low profit margins explain why there are not more suckler cows on lowland farms. The national herd is however on the increase, mainly owing to the rapid increase of cow numbers on the hill and upland areas.

Turning to fattening cattle, the main developments over the last decade have been concerned with a reduction in the average age at slaughter from over 24 months to 18–20 months.

Beef fattening is now generally practised under two main systems— intensive feeding on cereals and arable by-products coupled with all-year-round housing, or a system based on grass and conserved forage. The former is, quite naturally, found mainly on the larger arable farms, and in these cases it is uncommon to find a large sheep flock carried. Break crops such as peas, beans, sprouts or other vegetable crops are usually preferred to grass leys for rotational purposes.

Barley Beef enjoyed considerable popularity in the initial years after its introduction by Dr. Preston of the Rowett Research Institute. But after the first two years the system received a series of setbacks. Calf prices rose steeply from £15–£25 per head owing to the demand, pneumonia started to become troublesome in calves in badly designed houses, and there was a slight consumer resistance to the weaker flavoured beef that was produced. At this time many expensive buildings were erected and proved a financial liability when margins per head declined. More experienced advisers were only recommending farmers to go into barley beef if they had spare buildings and labour available coupled with a regular supply of cheap barley. The margin was generally estimated as equivalent to the calf subsidy, i.e. £10 per head. Since the end of 1967 the price of beef, due to increased consumer demand and to higher guaranteed prices, has risen by up to £3 per cwt with a resulting improvement in the sale price of the fat beast. Also scientists have now ironed out most of the problems associated with this form of beef production and a second small "boom" may develop in the next few years.

The forage-based beef production systems of grazing strong stores, and fattening suckled calves on silage and barley are both still quite profitable. The main skill involved is in the buying and selling of these animals rather than in their method of feeding.

A more recent development has been that of 18-month beef. Based on work by I.C.I. and others, the system is now fairly well understood and is producing Gross Margin figures equal to that of some of the cash root crops and superior to most cereal crops. Gross Margins of £40–£60 per acre are frequently reported in the farming press and it is likely that the system will enjoy increasing popularity in the next few years. With an 18-month beef system the animals are forced from birth to slaughter, and after weaning are intensively stocked on heavily fertilised grassland during the summer, followed by a winter ration of silage and barley. With stocking rates of up to two beasts per acre the capital investment again becomes a most important factor. As the second group of calves has to be purchased before the first group has been sold, the peak capital requirement may well be in excess of £100 per head—and if the stocking rate, including acreage for winter feed, is 1 acre per head, the capital investment is almost comparable to that of a dairy herd. The main advantage of the system is that it can yield a gross margin not much less than that of dairying but with much less labour required—particularly at peak periods in the arable farmer's year. It is also possible to work a 5- or 6-day week if the building is carefully designed.

On many grass farms a sheep flock is carried as a complement to cattle and also to act as scavengers. One of the oldest husbandry rules is that sheep should not graze the same field on successive years if at all possible because of worm problems, e.g. Nematodirus. Thus the sheep flock often graze a field in the first and third year and cattle in the second year, or alternatively, the field is cut for hay or silage in the second year. Mixed stocking with cattle and sheep coupled with a balanced fertiliser policy can often result in a marked improvement in pasture quality and productivity; hence the continued popularity of a mixed stocking system.

Table 12.6 shows that these beef and sheep systems are not producing a reasonable income except on the larger farms. The output per acre is

Table 12.6
Beef and sheep farms

Item	Units	Business size (standard man days)					Average of all farms
		275–599	600–1199	1200–1799	1800–2399	2400–4199	
Average acreage	Acres	170	316	615	*	*	264
Average labour required	SMD	440	848	1422	*	*	682
Distribution of holdings	%	40	44	12	(4)	*	—
Gross output per acre	£	22·7	24·9	22·7	*	*	23·2
Net output per acre	£	18·2	20·8	19·5	*	*	19·2
Net Farm Income per farm	£	1139	2130	3315	*	*	1686
Net Farm Income per acre	£	6·7	6·9	5·4	*	*	6·4
Manual labour of farmer and wife	£	680	638	562	*	*	661
Management and Investment Income per farm	£	459	1492	2753	*	*	1025
Management and Investment Income per acre	£	2·7	4·8	4·5	*	*	3·9
Tenant's capital per acre	£	34·7	36·5	33·8	*	*	35·5
Return on tenant's capital	%	7·8	13·1	13·3	*	*	10·9
Farmer's Income	£	681	1211	1630	*	*	951

* These figures are not available

generally too low to make these systems profitable on smaller acreages, and it is only farms over 300 acres which can afford to run cattle and sheep on slightly less intensive lines and go for a low cost low output policy. As with the dairy and sheep farms, the farmer is contributing a large proportion of the manual labour on the large farms, and it is only the two larger business size groups which generate acceptable levels of return on capital and Farmer's Income.

EXTENSIVE CEREAL FARMS

Extensive cereal farms are those on which the cereal crop constitutes over 75 per cent of the total labour requirement. The remaining labour is utilised by a wide variety of subsidiary enterprises such as pigs, barley beef, small sheep flocks, poultry or by break crops such as peas, beans, grass seeds, sugar-beet seeds or some vegetable production. The areas where this system is most frequently found are the Cotswolds, Wiltshire, Hampshire, parts of Northumberland and on the eastern counties of Scotland from Berwick to the Moray Firth. In East Anglia, potatoes or sugar-beet usually play a more prominent part; this is discussed in more detail in a later section.

Many of these extensive cereal farms are either owner-occupied or held on a very low-rent tenancy. In the former many farms changed hands in the period 1920–1945 at prices of £10–£30 per acre and large farming units were built up. This sector of the industry has enjoyed perhaps the most spectacular increase in land values to the point where a price of £300 per acre is often insufficient to purchase such a holding. Many owner-occupiers do not include a rent charge in their accounts, or base this rent figure on below-average land values; hence they are in a very favourable position when compared with a tenant paying £8–£10 per acre for a mixed or grassland farm.

The restrictions imposed on landlords by the various Agricultural Holding Acts mean that many tenant farmers on these larger farms have, until very recently, been paying rents of £3–£6 per acre. With land valued at only £250 per acre this represents a return of less than 3 per cent on the capital invested by the landlord. So the background to many of the extensive corn growing systems is one of low rents resulting in little incentive to intensify or little necessity to generate high profits before rent. Along with this has come the popular vogue for simplifi-

Table 12.7
Extensive cereal farms

Item	Units	Business size (standard man days)					Average of all farms
		275–599	600–1199	1200–1799	1800–2399	2400–4199	
Average acreage	Acres	160	307	498	694	880	328
Average labour required	SMD	475	872	1486	2024	2750	965
Distribution of holdings	%	22	44	18	9	7	—
Gross output per acre	£	39.0	39.4	39.0	38.5	42.8	39.5
Net output per acre	£	32.2	35.5	33.7	34.3	38.1	34.4
Net Farm Income per farm	£	888	2705	3712	5923	8328	2616
Net Farm Income per acre	£	5.5	8.8	7.5	8.5	9.5	8.0
Manual labour of farmer and wife	£	548	502	391	312	320	485
Management and Investment Income per farm	£	340	2203	3321	5611	8008	2131
Management and Investment Income per acre	£	2.1	7.2	6.7	9.1	9.2	6.5
Tenant's capital per acre	£	46.6	46.5	43.5	39.5	45.5	45.0
Return on tenant's capital	%	4.5	15.5	15.3	20.5	20.2	14.4
Farmer's Income	£	292	1545	1965	3704	5033	1423

cation and streamlining of the farm to a very small number of enterprises, even down to an all-corn policy in certain instances. The prices and yields of the cereal crops in the early 1960's gave a very useful margin per acre for a low labour and machinery input and therefore it is not surprising that there was an explosion in barley acreages and a considerable reduction of livestock carried and workers employed. The average size of farm for the different business groups is much higher than any other of the systems so far discussed (see Table 12.7). It is not uncommon to find farms of 600 acres employing only two men—that same farm probably having had a labour force of 15–20 before the 1939–45 war.

It is obvious, with such a large proportion of the farm in cereal crops, that many fields will be growing cereals over a long period, and in certain cases continuously. Not surprisingly there usually follows a drop in yield although this appears to reach a base level and does not decline beyond this for successive crops.

The decline in yield is a combination of three main factors: a marked increase in cereal diseases and weeds, a breakdown in soil structure, and a reduction in the natural fertility of the soil. As these points have been covered in detail in earlier chapters, it is sufficient to point out that if the Net Farm Income is to be maintained under an extensive cereal system, either

(a) costs have to be pruned to the absolute minimum to offset the falling income, or

(b) considerable extra work must be undertaken and cost incurred in stubble cultivation, weed control, extra fertiliser and purchase of more resistant varieties to maintain the level of income.

In the last three years the effects of diseases such as leaf blotch, rusts, smut and mildews, coupled with increased rents have seriously reduced the Net Farm Income of many of these extensive corn farms. It is likely that a greater reliance will be put on break crops or non-land-using intensive livestock enterprises to boost the Net Farm Income.

Table 12.7 shows that:

(a) Net Farm Incomes for all farm sizes are fairly standard at £8–£10 per acre;

(b) the farmer's manual labour rapidly reduces as the farm size increases;

(c) the tenant's capital per acre is also fairly constant at £45 per acre;

(d) the return on capital is generally quite satisfactory and varies from 15–20 per cent except on the smaller farms;

(e) the farms are able to generate a substantial income and hence a good standard of living for the majority of farmers—particularly as 80 per cent of the farms would produce a Farmer's Income of over £1500 per annum.

CASH CROPPING

The various break crops in extensive cereal systems play a relatively minor part in the farm economy. The acreages of these crops in the cash cropping group, particularly potatoes and sugar-beet, are such that they are of far greater importance. These two crops form the basis of the economy in certain areas such as East Anglia, Lincolnshire and eastern Scotland. Cereals are often grown to fill in the crop years between two crops of these cash roots. On these cash cropping farms there are often few or no livestock, with the exception of pigs and poultry. On some of the top-class soils a rotation of potatoes, wheat, barley, sugar-beet is quite common; i.e. 50 per cent of the farm is in high value cash root crops. These two crops tend to require labour at the same periods of the year, and consequently there is a tendency for vining peas, sprouts, celery, carrots or onions to replace one of these major crops. The predominance of sugar-beet in Suffolk and Norfolk, and of potatoes in Lincolnshire and the area surrounding the Wash, emphasises this point.

The most significant development in these cash cropping areas is that of contract farming. Peas, sprouts, carrots, raspberries, celery, etc. are all grown on contract to large freezing or canning firms—and in certain cases the manufacturer provides an extremely precise timetable for the farmer to follow and relieves him of some of his managerial problems. These high value crops have also resulted in the formation of farmers' co-operatives to market the produce, or more commonly machinery syndicates to share the very expensive machinery required for such tasks as pea vining, celery washing, sprout picking and multi-row sugar-beet lifting.

Many of these high value cash crops have heavy labour demands, hence the average labour force employed is above that of the extensive cereal growers. This can be seen from Table 12.8 in that 55 per cent of the farms were employing over 1200 SMD, i.e. more than four men. The techniques of production have greatly improved over the last decade

Table 12.8
Cash cropping farms

Item	Units	Business size (standard man days)					Average of all farms
		275–599	600–1199	1200–1799	1800–2399	2400–4199	
Average acreage	Acres	80	162	273	381	510	207
Average labour required	SMD	445	876	1433	2078	2922	1137
Distribution of holdings	%	15	30	24	12	19	—
Gross output per acre	£	60·0	63·9	60·6	65·0	66·5	63·4
Net output per acre	£	50·2	53·0	51·6	54·5	55·5	53·4
Net Farm Income per farm	£	1222	2342	3749	4678	6245	2795
Net Farm Income per acre	£	15·2	14·5	13·7	12·3	12·2	13·5
Manual labour of farmer and wife	£	664	576	521	387	315	553
Management and Investment Income per farm	£	558	1766	3228	4291	5930	2242
Management and Investment Income per acre	£	7·0	10·9	11·8	11·3	11·6	10·8
Tenant's capital per acre	£	58·5	64·9	54·9	59·0	54·6	58·4
Return on tenant's capital	%	12·0	16·8	21·5	19·1	21·3	18·5
Farmers Income	£	852	1496	2547	2861	4005	1824

with the rapid development of new machinery, varieties, herbicides and pesticides. Better organised marketing and a more standardised product have made the average returns from these cash crops very attractive. However, these averages can be very misleading in that most of these cash crops are high risk crops. The vagaries of the weather, disease incidence, yield and consumer demand result in very wide year-to-year variations which are not generally found in other sectors of the industry, e.g.

Potato sales vary from £100–£350 per acre
Sugar beet sales vary from £50–£150 per acre
Peas sales vary from £20–£150 per acre
Strawberries and raspberries sales vary from £100–£400 per acre

The average figures in Table 12.8 show that the wide variations in the different crop sales rarely all occur in the same year and that the cash cropping farms are the most profitable under present price levels. The Gross Output per acre is comparable to that of many of the dairy farms, and the Net Output is in fact higher than the dairy farms. The Net Farm Income for each group averaged £12–£16 per acre, and all but the smallest group left a reasonable figure for Management and Investment Income.

The farmer was able, as with the extensive cereal farms, to devote more of his time to managerial duties as the farm size increased. It is not surprising that with these specialist high value crops the investment in Tenants' Capital is quite high at £55–£65 per acre. On those farms on which processing or vegetable storage plants have been erected, the tenant's capital may well be considerably higher. The substantial levels of Management and Investment Income when related to the tenant's capital produce a rate of return (12–22 per cent—average 18·5 per cent) higher than that of any other land-using farming system, with the possible exception of the very efficient dairy unit.

It has been mentioned that grazing livestock play a very minor role in these cash cropping systems, but pig production is of much greater significance. In the major arable counties of Norfolk, Suffolk, Lincolnshire and Cambridgeshire, over 20 per cent of the pig population of England and Wales can be found. Much of this is the fattening of weaners brought in from the Midland counties, and also from local farmers; but the sow and gilt population is showing a steady increase. Future market prices for barley will however have a considerable effect on whether this trend continues. There are some dairy herds on these intensive cash cropping farms, but their number is slowly declining except for some very large farms, some of which are installing new units of 150 cows and above.

Capital investment and labour requirement per acre for these grass-using enterprises have increased with the rapid improvement in stocking rates, and unless the unit is of sufficient size to utilise a reasonable proportion of the rotation these enterprises only add to the manager's problems. Hence the preference for various break crops, e.g. peas, beans and oil-seed rape, which do not require any specialised machinery beyond that normally kept for cereal production.

MIXED FARMS

The final main group of land-using farming systems is that of the traditional mixed farm characteristic of most regions of Britain. The general pattern includes some permanent pasture and some 2- to 5-year leys, with the remainder of the farm (up to 50 per cent of the total acreage) under cultivation for cereals, feed roots and a limited acreage of potatoes or sugar-beet. This pattern follows closely the lines of good husbandry practised for the last 300 years. On few farms would cereals be grown for more than 4 years in succession on the same field, and the rotation is likely to be: wheat, barley, barley 2- to 5-year ley, potatoes, etc.

Emphasis is placed on soil structure and fertility and the livestock enterprise is to be expected to return much of the nutrients to the soil through farmyard manure. Several basic concepts underlie these mixed farming systems, and all are of great importance in determining the correct policy for a particular holding. A mixed farming policy should produce a more even labour demand throughout the year than a specialist system tending towards monoculture. A bad spell of weather or poor market conditions may be disastrous for one enterprise but beneficial to another. With a mixed system you can discount against risk to a certain extent. As mentioned above, better control can be exercised over soil structure and fertility, pest and disease incidence and other factors of long-term significance to the farm productivity. Much of the machinery is able to be used more widely during the year due to the absence of really specialised equipment. On some extensive cereal farms, combine harvesters costing £8,000-£10,000 may be employed for as little as 6 weeks per year. With a mixed system the flow of cash into and out of the business may be far less erratic and seasonal than with any other system except dairying. A simplified two- or three-enterprise system usually requires heavy capital investment in specialised buildings and machinery and a high degree of skill on the part of the workers. If these basic requirements are not available a mixed farming policy may well make the best use of those resources which are at the farmer's disposal. Finally many farmers and managers are not business-orientated and would not be capable of handling one or two large-scale specialised units, whereas the management problems posed by a number of smaller enterprises are well within their grasp.

It is not surprising that with the wide variation in topography, layout and soil types between farms, and also within farms, the advantages and disadvantages of a mixed farming policy are frequently discussed at farmers' meetings and conferences throughout Great Britain. In the late 1950's and early 1960's the trend was towards simpler systems; but due to the declining yields on some cereal farms, and recent improvements in margins of some of the livestock enterprises, there is now a trend to return to more traditional mixed farming systems.

The more significant factors in Table 12.9 are:

(1) The smaller mixed farms are finding great difficulty in generating a decent standard of living for the farmer, and are only yielding a low return on the capital he has invested.

(2) The farms of over 190 acres are much more economic: the Gross and Net Output figures are comparable to many of the other farming systems, and the Total Net Farm Income is more satisfactory.

(3) The Net Farm Income for all the farms in the group is fairly constant at £9–£12 per acre. The farmer is able to work less hours as the farm size increases and the average reward for his management and investment on the larger farms is constant at £8 per acre.

(4) The investment in Tenant's Capital is second only to that of the specialist dairy farms. This is mainly due to the large numbers of livestock carried.

(5) The return on capital for the farms over 150 acres average 12–17 per cent, just above the minimum that a farmer should be prepared to accept—particularly if his farm is financed with borrowed capital. If the farmer wishes to have a reasonable standard of

Table 12.9
Mixed farms

Item	Units	Business size (standard man days)					Average of all farms
		275–599	600–1199	1200–1799	1800–2399	2400–4199	
Average acreage	Acres	117	189	272	394	565	210
Average labour required	SMD	475	878	1502	2070	3006	1021
Distribution of holdings	%	15	31	23	13	18	—
Gross output per acre	£	43·7	52·4	68·4	64·3	64·3	56·6
Net output per acre	£	31·2	39·1	46·3	43·4	47·8	40·7
Net Farm Income per farm	£	1173	2170	3482	3989	5013	2272
Net Farm Income per acre	£	10·0	11·5	12·8	10·1	8·9	10·8
Manual labour of farmer and wife	£	635	618	530	530	250	585
Management and Investment Income per farm	£	538	1552	2952	3459	4763	1687
Management and Investment Income per acre	£	4·6	8·2	10·8	8·8	8·4	8·3
Tenant's capital per acre	£	47·4	57·2	64·6	70·3	64·2	59·1
Return on tenant's capital	%	9·7	14·4	16·7	12·5	13·1	14·0
Farmer's Income	£	730	1305	2077	2756	2081	1275

living, to be able to pay his income tax and to be able to repay his borrowing in under 10 years, then a minimum return of 15 per cent on his Tenant's Capital is required.

PIG AND POULTRY FARMS

The land is only of minor significance on the majority of pig and poultry farms. Most of the stock will be housed throughout the year, although the running of pregnant sows in paddocks, or fold-rearing and free-range poultry, are still quite common in certain areas. Any land on the larger farms not used directly by the livestock is usually used for feed production, mainly cereals, for home processing by the livestock. Fattening of store cattle may be practised on some of these farms, but the major emphasis will be on the intensive livestock enterprises.

Pig production

It is not surprising that with food being 70 per cent of the cost of pig production the fattening units are more common near the main arable and cereal-growing districts. The breeding units are more evenly spread throughout the country. Since the pig is basically a non-land using animal, a large percentage of the pig population is found on smaller farms, i.e. those farms which need an increase in their output but are severely limited by acreage. The average herd size is still only 12 sows in spite of the increasing number of units of 300–600 sows. Approximately 20 per cent of all the farms in England and Wales have a pig unit of some sort, with an average of 75 pigs per holding. In spite of the well-known fluctuations in returns and pig numbers (the Pig Cycle), the pig is still vitally important to the economy of many small farms. The main decision that any prospective pig producer has to resolve is, which line of production is he going to follow:

(a) Breeding herd,
(b) Fattening Herd,
(c) Breeding and fattening.

The recent development of weaner groups sponsored by farmers' groups or by feed merchants has produced a swing away from the combined breeding and fattening units, particularly on smaller farms. The high capital cost per fattening pig place (£10–£25) means that the investment in buildings can be too great on the smaller farms. Breeding stock, with the exception of the first 3 weeks after farrowing, do not require expensive environmental controlled housing; and in certain cases, notably the Roadnight System of outdoor sow-keeping, the capital cost of housing can be as low as £20 per sow.

The smaller farms tend to specialise in weaner production and to sell their weaners through a group to large fattening units. The availability of cereals on the larger farms and the desire to take a second "profit" on these crops is responsible for the rapid development of large fattening units. However, some of these farms, which do not have a contract with weaner groups or a feed manufacturer, are often faced with an irregular supply of weaners in the terms of price, quality and numbers. Hence many have decided to introduce their own breeding herds to regulate the supply position. The combination of breeding and fattening units on the same farm may also help control some pig diseases, e.g. virus pneumonia. The main attractions with any form of pig production are quick turnover of capital and very high outputs from a limited acreage. For these reasons many new entrants into agriculture are advised to start with their own pig unit on a small acreage rather than risk under-capitalising and thus under-exploiting a larger farm.

As feed and labour constitute up to 90 per cent of the cost of production, it is not surprising that these two factors receive most attention in efficiently managed units. It is essential to have a feed recording system which is quick and easy to work and understand, and which eliminates wastage wherever possible. The utilisation of labour by a pig enterprise is different to any other enterprise on a farm. The labour is usually employed solely for that unit and is rarely used on any other enterprises.

It is vitally important to plan the unit so that the man is in a position to run as many pigs as possible without outside assistance and without wasting too much of his time on non-productive operations. It is essential therefore to plan in terms of man units and to provide suitable facilities for the labour force to work as a self-contained self-supporting unit.

Poultry production

In poultry production, three main systems are in common use: laying hens, broilers and turkeys. Ducks and capons have recently been concentrated on to two or three very large farms and do not now play any significant part on the majority of poultry holdings, the numbers usually being restricted to those needed for consumption by the farmer and his immediate family and friends. The laying flocks for commercial egg production are gradually moving from free-range or deep-litter systems into some form of battery housing due to lower capital cost, space requirements and better labour utilisation. The breeding flocks for laying birds or broiler production are still basically run on a deep-litter system.

The recent rapid growth of very large 100,000 bird units owned by one or two private or public companies has produced many changes in the pattern of commercial egg production over the last decade. The market in eggs following the demise of the British Egg Marketing Board is likely to result in a fall in egg prices, and it may be increasingly difficult for the smaller producer to remain competitive. Flocks of under 5,000 birds may survive, as the units can be worked with part-time labour and the eggs sold at the farm gate or on a retail round for premium prices. Flocks of 5,000–15,000 however, with a permanent full-time labour force and with the majority of eggs being sold to wholesalers, may find the margin diminishes rapidly so that this size unit is unprofitable. Above 15,000 birds the economies of scale in food and chick purchasing should enable farmers to maintain their positions under a free market policy for eggs.

Hatching eggs with their extra price per dozen, although offset to a certain extent by low egg yields, should remain a reasonably profitable proposition for those farmers who hold contracts for this form of production.

Broiler production underwent a great increase in production in the late 1950's followed by an equally great slump through over-production and decreasing margins. In the last 3 or 4 years however, some degree of stability has returned to the industry. An excellent series of publicity campaigns, coupled with very attractive prices, has persuaded consumers to change their attitude to chicken meat. It is no longer regarded as a luxury, and is in fact now the cheapest meat on the market except for some of the poorer cuts of lamb. The development of processing and pre-packing for the supermarkets has also helped the sales of chicken, and poultry-meat is taking a larger proportion of the total meat market as shown in Table 12.10. Home poultry-meat production has increased from 70,000 tons in 1946–47 to 554,000 tons in 1969-70 and is continuing to increase with an annual increment of 40,000 tons. Broiler units are generally found on the larger arable farms and are employed as cereal processors and producers of dung for the maintenance of the cereal output.

The margin per bird varies from 2–5p; thus it is obvious that only the really large units are likely to leave a substantial profit—and profitable units of under 20,000 are now uncommon. The average age of slaughter is 9 weeks and allowing 3 weeks cleaning and disinfecting period at least four batches can be put through each house annually—so with housing costing 50–100p per bird housed the quick turnover produces an acceptable return on capital.

Turkey production has traditionally been one of the major sources of pin-money at Christmas time for the farmer's wife—but with the increasing capital investment in all forms of farm buildings, more farmers are interested in utilising these buildings more fully during the year. After the initial rearing stage a controlled environment is not essential for the fattening phase; hence many cattle courts, floor grain stores etc. are now used for a crop of turkeys during the summer months.

Table 12.10
Pattern of U.K. meat consumption

	1946/7	1963/64	1967/68	1969/70
Beef	50%	37%	33%	33%
Mutton and lamb	29%	18%	17%	16%
Pork, bacon and ham	15%	34%	35%	35%
Poultry-meat	6%	11%	15%	16%
	100	100	100	100
Total supplies (in 1,000 tons)	1,894	3,323	3,336	3,591

Table 12.11 shows that the very small producer on an average acreage of 29 acres is hard-pressed to make a decent living. The very low Management and Investment Income suggests that the farmer is working for a below-average wage and would, in terms of financial income, be advised to give up his farm unless he can produce above-average results.

The results of the second group, although showing some improvement, would also cast some doubts on the viability of these holdings. With the 1200–1800 SMD Group the financial results are much more attractive and the degree of intensification is reflected in the very high Gross Output per acre. The Net Output reflects the reliance on purchase feeds. The remaining figures for this group are all quite satisfactory and the very satisfactory return on capital at 23·8 per cent is above the minimum of 15 per cent suggested in an earlier section. This group are basically pig or poultry producers with a limited amount of cereals—probably grown and harvested by contractors.

The over-1800 SMD group were in a position owing to their larger acreage to be able to mechanise the production of the cereal crops and to produce both intensive livestock and cereals efficiently and profitably.

HORTICULTURAL HOLDINGS

Very small holdings of less than 10 acres have not been considered, as small-scale market gardening does not come within the definition of a farming system. With the very high labour requirements of vegetable and fruit crops, few farm scale horticultural units require less than 600 SMD.

The remaining four business size groups in Table 12.12 are all producing good results—particularly on a per-acre basis. These are of course influenced by the land utilised for horticulture being potentially capable of much higher outputs than the land types found in all the preceding groups. Horticultural land is sold for prices in excess of £500 per acre, hence it is perhaps of little interest to study these per-acre figures.

The Net Farm Income and Management and Investment Income are sufficient to provide a moderate standard of living for the farmer.

The majority of vegetable crops are very dependent on soil type and climatic conditions, and so the variation in annual returns is likely to be greater than for more extensive farming systems. Much has been written of the need for better marketing of farm produce and of the reduction in the number of links in the chain between grower and consumer. The grower's cost for many vegetable crops is often as low as 25 per cent of the price paid by the consumer. It is clear, therefore, that for these horticultural holdings there is considerable scope for

Table 12.1
Pig and poultry holdings

Item	Units	Business size (standard man days)					Average of all farms
		275–599	600–1199	1200–1799	1800–2399	2400–4199	
Average acreage	Acres	40	77	120	266	210	79
Average labour required	SMD	384	895	1422	1992	3145	947
Distribution of holdings	%	18	36	15	15	16	—
Gross output per acre	£	166·1	209·0	184·0	127·1	276·5	197·7
Net output per acre	£	68·1	87·5	88·0	67·4	113·8	85·4
Net Farm Income per farm	£	908	1863	4094	6728	7117	2217
Net Farm Income per acre	£	22·7	24·2	34·1	24·9	33·9	28·0
Manual labour of farmer and wife	£	594	637	597	579	525	603
Management and Investment Income per farm	£	314	1226	3479	6149	6592	1614
Management and Investment Income per acre	£	7·9	15·9	29·1	23·1	31·4	20·4
Tenant's capital per acre	£	99·1	158·5	122·3	100·1	142·5	129·9
Return on tenant's capital	%	7·9	10·0	23·8	23·1	22·1	15·7
Farmer's Income	£	591	874	2920	4607	4599	1386

Table 12.12
Horticultural holdings

Item	Units	Business size (standard man days)					Average of all farms
		275–599	600–1199	1200–1799	1800–2399	2400–4199	
Average acreage	Acres	11	28	39	76	60	28
Average labour required	SMD	484	945	1493	1943	3013	1087
Distribution of holdings	%	9	38	26	11	16	—
Gross output per acre	£	295·8	240·0	328·8	171·0	329·0	274·0
Net output per acre	£	266·5	209·2	272·8	158·5	298·5	246·5
Net Farm Income per farm	£	943	1929	3102	3311	3567	1930
Net Farm Income per acre	£	85·8	68·8	79·5	43·6	59·5	68·9
Manual labour of farmer and wife	£	717	707	627	494	505	664
Management and Investment Income per farm	£	226	1222	2475	2817	3062	1266
Management and Investment Income per acre	£	20·5	43·7	63·4	37·1	51·0	45·2
Tenant's capital per acre	£	*	*	*	*	*	*
Return on tenant's capital	%	*	*	*	*	*	*
Farmers Income	£	*	*	*	*	*	*

* These figures are not available

improvement in farm profits by more efficient marketing. This generally requires the provision of suitable grading and packing facilities on a co-operative basis. The advantages of this and other measures have been discussed in an earlier chapter.

EXCHEQUER SUPPORT TO AGRICULTURE

In an industrial society the government is faced with the choice between a free market in food or a controlled subsidised policy. A free market means that the prices the farmer receives for his products are determined by supply and demand, and the farmer is free to produce any product he may wish. This free market generally results in high food prices with consequent high industrial wage levels. The alternative is for the government to control the prices of farm products and to compensate the farmer for any loss of income due to these lower prices. This results in a lower direct cost of living, and hence lower industrial wages, but does impose extra taxation on the country as a whole. In this case the farmer should be receiving a fair price for his product, but this price will be made up of the controlled market price plus a government subsidy. Therefore in a free market the consumer pays the market value to the farmer. In a controlled market the consumer pays a lower price to the farmer, *some* consumers pay tax to the government, and the government pays a subsidy to the farmer.

As 97 out of every 100 workers in Great Britain are employed outside agriculture, low industrial wage levels are attractive to the national economy. Successive governments since the war have therefore adopted a controlled subsidised market for food products. They have agreed to compensate farmers for any loss of income caused by this policy when compared with a free market policy. This compensation is paid in the form of guaranteed prices and deficiency payments. The former are equivalent to the controlled price, and the latter is the estimated difference between controlled and free market prices.

Great Britain is still a net importer of food despite rapid increases in productivity (Agriculture 6 per cent per annum, and Industry 3 per cent per annum). This is partly due to the inability to grow certain products, notably hard wheat for bread, cane sugar, tea, coffee, etc.

Secondly, a large proportion of our industrial exports go to the less industrially developed countries. These countries can only raise currency to pay for these goods by the export of food and raw materials. Thus to sell exports we are asked to buy imports in the form of food products. These reciprocal trade arrangements result in increases in supplies of food products on to the home market which, due to the fairly constant demand for most foods, bring about falls in the prices received by the home producer. The present support system is designed to take this depressing effect on prices into account.

The encouragement of improvements in efficiency in the home industry and the striking of a reasonable balance between imported and home-grown foods is achieved by the government being able to adjust the prices received by the farmer through the Annual Farm Price Review. They are also empowered to offer various production grants and subsidies to encourage production of those commodities which it is in the nation's interest to produce at home. This policy was laid down in the 1948 Agriculture Act. If the income of the farming industry is to be maintained at a satisfactory level within this structure of controlled prices for food products, it is clear that some assistance is required in keeping the increase in costs to a minimum. The production grants such as the fertiliser subsidy, drainage grants and winter keep grants are designed for this purpose, and also to encourage further capital improvements and modernisation of agricultural holdings.

Summarising, the present support system

(a) compensates farmers for the loss in income caused by a cheap food policy linked to a competitive industrial sector,

(b) compensates farmers for the loss in income caused by imports of food obligatory under reciprocal trade arrangements,

(c) helps to reduce costs of production, as the farmer, unlike other manufacturers, is unable to recover his increased costs by raising prices to the consumer,

(d) encourages the modernisation and improvement of holdings which help the industry to maintain a growth rate of twice the national average for other industries,

(e) encourages changes in the pattern of production by price manipulation so that farmers can produce that proportion of the total food requirements which it is in the nation's interest to produce.

The average cost of exchequer support for the last ten years has been £250–£300,000,000 per annum, of which approximately one-half has been in the form of guaranteed prices, and one-half in production grants. At the time of writing the Conservative Government is planning an unsubsidised free market in food with protection for the home agricultural industry through import controls and levies, and this may be preferable to a support system. It is probable that food prices to the consumer will rise by 8–10 per cent on average. It is also probable that wage rates in the industrial sector will rise too, making our industrial goods less competitive. The farmer will, in theory, receive a fair price directly from all consumers, whereas, until now, he received this price through a controlled market system plus exchequer support, i.e. indirectly from some consumers via taxation. In practice it may be found that the demand for food is so inelastic that, if prices rise, the consumer may increase purchasing certain foods, e.g. bread and potatoes, and reduce the purchasing of others, e.g. milk, eggs, meat and vegetables.

FUTURE DEVELOPMENTS

The farming industry has undergone considerable change since the 1939–45 war. The advent of new chemicals and cultivation techniques coupled with improved varieties has made monocultural cereal systems possible. Breeding, feeding, and housing of livestock have shown equally dramatic improvements. These improvements are likely to continue and possibly accelerate in the future, and will obviously alter the farming systems discussed in the previous sections. In farm business management the major developments are likely to follow two lines: planning and marketing.

Farm planning

Due to the excellent work of the Advisory Services and the University Agricultural Economics Departments, farmers and managers are becoming increasingly aware of the advantages of using efficient planning and control techniques. With capital becoming more difficult to obtain, with rising rents and land prices, and with a general increase in the cost of the basic raw materials, farmers must be more business-minded. Much wider use of sophisticated planning techniques based on the Gross Margin system is likely to become generally accepted. The preparation of an annual budget of income and expenditure as a Cash Flow Statement is already very desirable for effective management, and may be a condition of future bank lending. Budgetary control systems coupled with an accounting system which can provide financial data monthly or at a minimum quarterly will become more common, particularly if the farmer is prepared to treat his accountant as an adviser rather than a person divorced from the farm business. These developments are now receiving excellent support and promotion from the Farm Management Association and by various commercial firms.

Marketing

Alongside the development in management techniques will come an increasing awareness that it is not enough to produce a crop or animal; it must also be marketed well if maximum profit is to accrue. In the last ten years the growth of co-operatives and farmer's groups has been comparatively slow in spite of considerable enthusiasm of those farmers who have joined. Many farmers are reluctant to give up their independence and accept the rigid rules of conduct that are essential for successful groups or co-operatives. Vegetable growers have shown what can be achieved and it is likely that many smaller producers will have to pool their resources if they are to survive in future. The grants available from the Central Council for Agricultural and Horticultural Co-operation may accelerate this movement, but unfortunately it is usually larger producers who take most advantage from these schemes.

A second development is likely to be the further expansion of crops and livestock grown on contract to a freezer, packer or canner. This already happens for many vegetable crops, sugar-beet, raspberries, oil-seed rape, seed corn, sprouts, peas, etc. and has recently been introduced for potatoes for chipping, canning and crisping. On the livestock side broilers, turkeys, weaner pigs, hybrid gilts and dairy replacement heifers are often produced on contract and new schemes are continually reported. For certain commodities, notably freezing peas, brussels sprouts, and broiler chickens, the grower is given a precise list of instructions from the processor and many of the husbandry decisions are taken out of his hands. This allows him to concentrate on cultivations and also on his other farming activities. In other cases, particularly with contract egg production schemes, the feed company or poultry supplier provides the birds, feed, veterinary products, etc., and pays the farmer a weekly or monthly sum for tending the flock and for the use of buildings and equipment. The farmer is thus virtually an employee of the company. Such schemes are attractive when the farmer's supply of capital is limited or where he is not prepared to stand the risk of losing money on an intensive livestock enterprise.

The final significant development which is already gathering momentum is the establishment of processing plants on the larger farms. Plants such as washing and freezing equipment, cold storage facilities and packaging machinery will enable the grower, or groups of growers, to see a graded and selected quality product direct to the supermarkets, thus eliminating several intermediate links in the grower-consumer chain.

The above possible future developments in Farm Planning and Marketing show that the management of the various farming systems is likely to pose many interesting problems. As these developments are taking place, new husbandry techniques will probably emerge and result in many variations of the existing farming systems. It will be up to the efficient manager to capitalise on these.

LIST OF BOOKS AND REFERENCES

Barnard, C. J., Halley, R. J., and Scott, A. H.: *Milk Production*. Butterworth (1970)
Barron, N.: *The Dairy Farmer's Veterinary Book*. Farming Press (1969)
Barron, N.: *The Pig Farmer's Veterinary Book*, 5th Edition. Farming Press (1970)
Barron, N.: *British Farmer's Veterinary Book*. Farming Press (1970)
Bowden, W. E.: *Beef Breeding Production and Marketing*. Land Books (1962)
Carter, A. R.: *Dutch Lights for Growers and Gardeners*. Vinton (1956)
Cooke, G. W.: *The Control of Soil Fertility*. Crosby Lockwood, London (1966)
Cooper, M. McG.: *Grass Farming*. Farming Press (1967)
Cooper, M. McG. and Thomas, R. J.: *Profitable Sheep Farming*. Farming Press (1965)
Culpin, C.: *Profitable Farm Mechanization*. Crosby Lockwood, London (1968)
Culpin, C.: *Farm Machinery*, 8th Edition. Crosby Lockwood, London (1969)
Davidson, H. R. and Coey, W. E.: *The Production and Marketing of Pigs*, 3rd Edition. Longmans (1964)
Dexter, K. and Barber, D.: *Farming for Profits*, 2nd Edition. Iliffe Books (1967)
Donaldson, F., Donaldson, J. G. S. and Barber, D.: *Farming in Britain Today*. Allen Lane, The Penguin Press (1969)
Duckham, A. N. and Masefield, G. B.: *Farming Systems of the World*. Chatto and Windus (1970)
Fryer, J. D., Evans, S. A. and Makepeace, R. J.: *Weed Control Handbooks*. Vol.1, *Principles*, 5th Edition, Blackwell (1968); Vol.2, *Recommendations*, 6th Edition, Blackwell (1970)
Garner, Frank H. and Norman, R. G.: *British Dairying*, 2nd Edition, Longmans Green (1972)
Halnan, E. T., Garner, Frank H. and Eden, A.: *Principles and Practice of Feeding Farm Animals*, 5th Edition, Estates Gazette (1966).
Hayhurst, J.: *Smallholder Encyclopaedia*. Pearson (1962)

Henderson, F.: *Build Your own Farm Buildings*, 4th Edition. Farming Press (1971)
Hutchinson, Sir J.: *Population and Food Supply*. Cambridge Univ. Press (1969)
Johnson, G.: *Profitable Pig Farming*, 3rd Edition. Farming Press (1968)
Luscombe, J.: *Pig Husbandry*, 3rd Edition. Farming Press (1970)
Laverton, Sylvia.: *Irrigation: Its Profitable Use for Agricultural and Horticultural Crops.* Oxford Univ. Press (1964)
Lockhart, J. A. R. and Wiseman, A. J. L.: *Introduction to Crop Husbandry*. Pergamon (1966)
McDonald, P., Edwards, R. A. and Greenhalgh, J. F. D. *Animal Nutrition.* Oliver and Boyd (1966)
Murdoch, J.: *Making and Feeding Silage.* Farming Press (1961)
Nix, J. S.: *Farm Management Pocket Book*, 3rd Edition. Wye College (1969)
Park, R. D., Coutts, L. and Hodgkiss, P. J.: *Animal Husbandry.* Oxford Univ. Press (1961)
Park, R. D., Harris, A. G. and Jones, T.: *Crop Husbandry.* Oxford Univ. Press (1961)
Preston, T. R. and Willis, M. B.: *Intensive Beef Production.* Pergamon (1970)
Payne, W. J. A.: *Cattle Production In The Tropics*, Vol.1. Longmans (1971)
Roy, J. H. B.: *The Calf*, 3rd Edition, Vol.1 and 2. Iliffe Books (1970)
Russell, K.: *The Principles of Dairy Farming*, 5th Edition, revised. Farming Press (1969)
Russell, K.: *The Herdsman's Book*, 3rd Edition, revised by S. Williams. Farming Press (1969)
Sainsbury, D.: *Animal Health and Housing.* Baillière, Tindell and Cassell (1967)
Sainsbury, D.: *Pig Housing*, 2nd Edition (1970)
Salter, P. J. and Goode, J. E.: *Crop Responses to Water at Different Stages of Growth.* Commonwealth Agricultural Bureau (1967)
Sayce, R. B.: *Farm Buildings.* Estates Gazette (1966)
Shewell Cooper, W. E.: *The Complete Vegetable Grower*, 2nd Edition. Faber (1966)
Spedding, C. R. W.: *Sheep Production and Grazing Management.* Baillière, Tindall and Cassell (1970)
Sturrock, F. G.: *Report On Farming, 1956-57.* Cambridge Univ. Farm Economics Branch Report No. 46 (1957)

Thompson, C. R.: *Pruning of Apples and Pears by Renewal Methods.* Faber (1949)

T.V. Vet, The: *T.V. Vet Book For Stock Farmers* No. 2. Farming Press (1965)

T.V. Vet, The: *T.V. Vet Book For Stock Farmers* No. 1, revised edition. Farming Press (1970)

T.V. Vet, The: *T.V. Vet Book For Pig Farmers.* Farming Press (1967)

The following are also useful publications:

Farm Electrification Handbooks: No. 9, Automatic Feeding (1964); No. 13, Grain Drying (1968); No. 15, Green Crop Drying (1967). The Electricity Council (E.D.A. Division), London.

The £.s.d. of Mechanization. Series of articles in *Farm Mechanization*, Vol. 18 (1966) Farm Journals Ltd., London.

The publications of the Ministry of Agriculture, Fisheries and Food should also be studied, together with reports from university and college departments too numerous to mention.

See also the Selected Bibliography of Modern Works on Agricultural History on pages 15 and 16 of this volume; reports on the Rothamsted Ley-Arable Experiment referred to at the end of Chapter 5 (page 93); and the notes at the end of Chapter 8 (page 147), dealing with further reading on fruit, flowers and vegetables.

INDEX

Acidity (soil), 19, 23
Agrarian history, 1–17
Agricultural Botany, Inst. of, 158
Agricultural Holding Acts, 219
Agricultural and horticultural co-operation, 238
Agricultural revolution, 7
Agricultural shows, 150
Agriculture Act (1920), 13
 (1947), 14
 (1948), 236
 (1957), 15
Agriculture, Central Council for, 238
All-arable systems, 35–58
 extensive, 43
 intensive, 56–8
Alternate husbandry system, 72–4
Amalgamations, 27
Annual Farm Price Review, 236
Apples, 146
Arable and fruit, etc., 142–5
Arable and sheep flock, 92
Arable tillage (maps), 36
Arminatriazole, 53
Automation, 198

Bacon and ham production, T. p. 14
Barley, 19, 23, 24, 38, 54, 85, 160, T. p. 14
 tillage (map), 38
Beans, 19, 23, 24
 field, 54
Beef (bull), 114, 116
 cattle, 22, 91–2, 216
Beef Recording Assn., 151, 215
Beef and sheep, economics, 215–18, T. p. 216, 218
Beef: U.K. consumption, T. p. 231
Beef and veal production, T. p. 14
Bibliographies, 15–16, 93, 147, 240
Blackface sheep, 94, 98
Blackgrass, 51
Bracken, 70
Break crops, 54, 165, 216
Breeding, artificial insemination, 150–3
 'line', 154, 156
 livestock, 148–56
 pedigree, 149, 153–6
 plants, 157–67
 for shows, 149
Brombell Report, 104, 106
British Seed Trade, 158

British Sugar Subsidy Act, 13
Brussels sprouts, 35, 56
Bulb growing, 136, 146
Butter production, T. p. 14

Cabbage, 146
Calves, intensive units, 109, 112–13, 114
Capital, 32–3, 201–2
Cappelle-Desprez, 46, 48
Carrots, 19, 56, 143, 146
Cash crops, 35, 222–5
Cattle: Aberdeen-Angus, 103
 Beef Shorthorn, 103
 buildings for, 104–5
 dairy cows, 110–11
 fattening, 87–9, 90–2
 intensive, 109, 114–16
Cattle, investment per acre, T. p. 216
 Scottish, 10
 Welsh, 11
 winter feeding, 102–3
Cauliflowers, 146
Celery, 21, 56, 143, 146
Cereals, 26, 35, 43, 55, 159–62
 extensive, economics of, 219–25, T. p. 220
 cyst eelworm, 47–8, 50, 159
 diseases, 159–61
 harvesting, mechanical, 181–2
Chalk soil, 18, 20
Cheese production, T. p. 14
Cherries, 146
Cheviot sheep, 94–5
Clay soil, 18, 19
 mixing, 23
Climate, 25
Clover, 5, 66–8, 164
CMPP, 50
Coleseed, 6
Conservation (fodder), 74, 81
Contract farming, 222
Contractors (vegetable crops), 145
Copyhold, 3, 4
Corn Laws, 8
Corn Production Act 1917, 13
Couch grass, 52
County executives, 12
Crop rotations, 57–8
 spraying (mechanical), 180
Cucumbers, 146
Cultivators (mechanical), 174–6

243

244 MODERN BRITISH FARMING SYSTEMS

Currants, 146
Cut flowers, 146

Dalapon, 53
D (2, 4-D), 50
DB, 51
Dairy and arable, 72, 87–90
  and cereal, economics, 208–9
  cows, 110–11
  farms, specialist, economics of, 205–8
  herds, 26
  imports, 9
  intensive units, 109
Dairying, 62–8
  fattening and arable, 72, 87–90
Demesnes, 2–3
Depressions, agricultural, 8
Derbyshire Gritstone sheep, 95
Diseases of crops, 158–65
DNOC, 51
Domesday Book, 2
Drainage, 9, 21
Drilling, cereal, mechanised, 179–80
'Dumping', 35
Dung, 64–5, 85–6, 106, 108

Economics, 31–2, 35, 56–7, 101, 200–39
Eelworm, 54, 56, 58, 158
Egg Marketing Board, 230
  production, T. p. 14
Enclosures, 4–7, 10
English Agricultural Society, 8
EPTC, 53
Equipment sharing, 27
Ewes, breeding flock, 116–18
Eyespot, 46–7, 50

'Factory' farming, 30
Farm income, 201
Farm Management Assn., 238
Farming systems, 17–34
Farms, sizes of, 17
Fattening and arable farms, 72
February Price Review, 34
Feeding, mechanical, 197–8
Fenlands, 18, 21, 22, 56
Fertilisers, 63, 64
  mechanical spreading, 176–9
Flower crops, 124, 138
  cultivation, 128
  fruit and vegetable farming, 124–47
  future prospects, 145–7
  growing specialists, 137
  labour, 127
  manuring for, 129–31
Food Production Department, 12
Forestry Commission, 213
  investment groups, 213
'Four course', 7
Free Trade, 8, 10, 13
Frosts, snow, 25–6

Fruit cultivation, 128
  growing, mixed, 137–8
  manuring, 129–31
  production problems, 140–1
  small, labour needs, 127
  specialist growers, 137
  spraying, 131–3
FYM, 129

Genetics, 154
Glasshouses, 125, 136
Gooseberries, 146
Government aids, grants, 104, 235–7
Government policy, 33–4
Grading (fruit), 133
Grain, chemical treatment, 185
  chilling (mechanical), 184–5
  drying, 183
  imports, 9
  storage (mechanical), 182–4
Grass, 19, 21, 23, 26, 54
  all-grass systems, 59–71
  grass drying, 81–2
  (mechanical), 195
Grasses, breeding of, 163–4
Grasslands, for fattening, 62
  for feeding, rearing, 69
  high quality, 60–1
  hill farms, 69–70
  low quality, 61–2
  permanent, 59–60
  seed mixtures, 67–8
Gravel soils, 18, 19
Grazing systems, 63–4
Green crop drying, 81–2

Hand feeding, 26
Harrows, mechanised, 175–6
Hay, haymaking, 79–80
  baling (mechanical), 194
  barns, 79
  costs, 80
  drying (mechanical), 194–5
  Dutch, 80
  mechanised, 193–4
  preservatives, 80
  quick making, 79
  tripoding, 80
Herbicides, 43, 53
Herdwick sheep, 95
'High farming', 8, 9, 11
Hill Cow Subsidy, 213
Hill Sheep Subsidy, 213
Hill-upland farms, 69–71, 94–103
History, mediaeval, 1–4
  16th–20th century, 4–15
Home production, T. p. 14
Hops, 35
Horticultural holdings, economics, 232–4
Horticulture, acres under (map), 42
Hybridization, cattle, 155

INDEX

Hybridization, *continued*
  cereals, 160–2
  pigs, 155–6
  poultry, 156
  sheep, 155–6

Improvers, Society of, 11
Income, of farm, 201–2
  of management, investment, 201, 224
Intensive livestock systems, 104–23
  units, 109, 120–3
Intercropping, on fruit, etc., farms, 138–9
Irrigation, 26, 27, 125

Jeffes' System, 55

Kale, 23, 24, 83–4, 164

Labour, costs, 43
  on fruit, etc., 126–7, 140
  gangs, 28–9
  housing, 29
  output, 30
  relief workers, 28–9
  requirements, 203–4
  supply of, 27
  wages, 29–30
Lambing, 96
Lambs, intensive units, 117–18
Land, price of, 41
Leaf diseases, 48, 49
Leaseholds, 3, 4
Leeks, 146
Lettuce, 146
Leys, 21, 23, 24, 65, 67, 73–4
Loam, 18, 20
Lonk sheep, 95
Lucerne, 6, 20, 23, 24

Maize, 54, 160, 161
Management income, 224
Mangolds, 84–5
Manorial system, 1–4
Manures, 20–5, 27–8, 85–6
Manuring on fruit, etc., 129–31, 139
Maris Badger barley, 158
Market gardening, 136, 232
Marketing, costs, 135
  display, 134–5
  fruit, etc., 133–4, 144
  future prospects, 238
  intelligence, 134
  wrappings for, 135
MCPA, 50, 51
Meadow fescue, 68
Meat imports, 9
Meat and Livestock Commission, 152
Mechanisation, 27, 28, 43, 168–99
  on fruit, etc., 128
  in milking, 198
  objectives, 168, 169
Mildew (cereal), 159

Milk Marketing Board, 13, 81, 151, 152
Milking, 29
Ministry of Agriculture, 73
  Report (1965), 104–5
Mixed arable-grass, 72–93
Mixed farms, economics of, 225–7
Modernisation, 18
Mountain rocks, 18, 22
Mutton-lamb consumption, T. p. 231
  production, T. p. 14

National Farmers' Union, 34
Nematodirus, 217
Nursery stock (horticultural), 146

Oats, 19–21, 23, 24, 54
  acres under, 39
  production, T. p. 14
  stem eelworms, 50
  tillage (map), 39
  wild, 52
Onions, 143, 146
Output, gross and net, 200

Paddock grazing, 23, 24
Paraquat, 45, 49, 53, 55
Parsnips, 143
Pears, 146
Peas, 19, 26, 35, 54, 56, 127, 166, 224
Peat land, 18, 21
Performance testing, 151
Personal interest, 32
Pests, 43–6
  on fruit, etc., 131–2
PH tolerances, T. p. 20, T. p. 23
Pig Board, 13
Pig Industry Development Authority, 151–2
Pigs: economics, 224, 228–9, T. p. 233
  fattening, 89–90, 119–20
  intensive units, 107–8, 109, 118–19
Planning, future, 237
Plant Varieties Act, 157
Plant Breeding Station, Cambridge, 159
Ploughs, mechanical, 173–4
Plums, 146
Population growth, 2, 7, 9
Pork, bacon, home consumption, T. p. 231
  production, T. p. 14
Potato Marketing Board, 13
Potatoes, 19–21, 26, 56, 84, 143, 165, T. p. 23, T. p. 24, T. p. 143
  acres under, 40
  economics, 224
  labour required, 127
  mechanised harvesting, 191–3
  mechanised planting, 190–1
  production, T. p. 14
  tillage (map), 40
Poultry, 90
  economics, 229–31

Poultry, *continued*
  intensive farming, 105, 107
  intensive units, 109, 120
  meat, T. p. 231
Powdery mildew, 49

Rackrenting, 4
Rainfall, 26
Rape, 6, 54
Refrigeration (fruit, etc.), 31
River Pollution Acts, 86
Roadnight System, 229
Root crop breeding, 164
Root crops, 35
Rotation, cereal and vegetable, 143–4
Rothampsted, 93
Rothwell Perdix wheat, 158
Rough Fell sheep, 95
Rye, 20, 163
Ryegrass, 60, 61, 66–8

Sainfoin, 5
Sand soil, 18, 19
Seed mixtures (grass), 67–8
Sheep, 22
  breeding, 98–101
  dipping, 97
  economics, 101–2, 211–14
  farming (hill and upland), 94–100
  farming (Scottish), 11
  fattening, 89
  intensive unit breeding, 116–17
  intensive units, 109
  sales, 97
  systems, T. p. 212, 214
Scotland, 10, 61, 94, 219, 222
Shepherd's Year, 95–7
Silage, 67, 74–9, 81, 85, 88, 111
  making, mechanised, 196–7
Silos, 183–4
Soils: chalks, 20–1
  clays, 19
  exhaustion of, 3
  fens, 21
  loams, 20
  mixing of, 23
  mountain rocks, 22
  peats, 21
  sand and gravel, 19–20
Slurry, 86–7, 111
Spraying, fruit flower and vegetable, 131–3
Steam power, 9
Straw balancer system, 82–3
Strawberries, 146
Subsidies, 235–7

Sugar beet, 13, 54, 56–7, 84, 164, 224, T. p. 23, 24, 41, 143
  cultivation, mechanised, 186–9
  harvesting, mechanised, 188–9
  tillage (map), 41
Swaledale sheep, 95
Swedes, 85, 164–5, T. p. 23, 24

Take-all, 44–7, 50, 55
TBA, 51
TCA, 53
Tenants' capital, 224
Tillage, acres under, 36
Timothy, 68
Tomatoes, 146
Top fruit trees, 138
Tractors, 170–3, 197–9
Turkeys, 107, 231
Turnips, 85
Turnover, 33

University Agric. Economics Depts., 237
'Up and down' husbandry, 5

Vegetables: crops, by districts, 124
  cultivation, 128
  economics, 232–3
  grading, 126
  integration with arable, 142–4
  labour, 126
  marketing, 127
  packing, 126
  specialist production, 141–2

Wages Board, 29
War 1914–18, 12
  1939–45, 13
Wart disease, 158
Weeds: broad-leaved, 50–1
  control of, 49–50
  on fruit, etc., farms, 132
  in grass, 52
Welsh Mountain sheep, 95
Wheat, 19, 54, 84, 143
  acres under, 37
  Act, 13
  labour requirement, 127
  production, T. p. 14
  winter, 23, 24
Wool, 98

Yellow Rust, 48–9, 159
Young Farmers' Clubs, 150

'Zero' grazing, 27, 30, 166

387